IET HISTORY OF TECHNOLOGY SERIES 36

Oliver Heaviside
Maverick mastermind
of electricity

Other volumes in this series:

Oliver Heaviside
Maverick mastermind
of electricity

Basil Mahon

The Institution of Engineering and Technology

Published by The Institution of Engineering and Technology, London, United Kingdom

© 2009 The Institution of Engineering and Technology

First published 2009

The Institution of Engineering and Technology
Michael Faraday House
Six Hills Way, Stevenage
Herts, SG1 2AY, United Kingdom

www.theiet.org

British Library Cataloguing in Publication Data
A catalogue record for this product is available from the British Library

ISBN 978-0-86341-965-2

Typeset in India by Newgen Imaging Systems (P) Ltd, Chennai

To Emily

Contents

Figures

Preface

I first came across the name Oliver Heaviside when I was a student of electrical engineering. My interest was aroused not by the lectures, where his name rarely surfaced, but by reading popular accounts of physical science. He seemed to have a strange hold over the writers. Every mention of his name carried a certain frisson – a hint of a great genius at work behind the scenes.

Over the years I built a mental picture of Heaviside from snippets in books and articles. It was a fuzzy picture, composed, as I found later, of truths, half-truths and outright fantasies: according to legend he was self-educated; he lost his job as a telegrapher because he was deaf; his work was suppressed by the authorities; he was swindled out of a fortune; he discovered relativity before Einstein; he lived his last years as an old-world hermit with granite blocks for furniture... A beguiling character, but shrouded in mystery. Only after retiring from regular work did I begin to discover the real man, and he turned out to be more fascinating than any of the myths.

While researching Heaviside I recalled an incident from student days. In the course of a laboratory experiment we had to work out the theoretical response of our circuit to a voltage pulse. The demonstrator went quietly to a corner of the blackboard, wrote an equation containing the unusual symbol p, had the answer in a trice and gave us an arch look, as though to say 'I know something you don't'. Indeed he did – he had solved the problem using Heaviside's operational calculus. But what he may not have known was that almost every part of the electrical engineering course bore Heaviside's distinctive stamp. General circuit theory, the theory of transmission lines and wave propagation, vector analysis, even the four famous Maxwell's equations; all had flowed from the pen of this amazing man.

In a way, Heaviside has fallen victim to his own (delayed) success. His great innovations, considered outrageous during his lifetime, have now become so familiar that almost nobody wonders where they came from. By the same token, our great debt to him is rarely acknowledged.

His writings reveal a personality like no other, and are laced with deliciously irreverent humour. A distinguished professor told me how, when reading on a train, he couldn't stop himself laughing so loudly that other passengers began to move to the next compartment. The book on his lap was Heaviside's treatise *Electromagnetic Theory*.

Anyone who writes about Heaviside for the general reader faces the essential but difficult task of conveying the essence of his remarkable mathematics in everyday language. I hope I have done this adequately. Those who want to go deeper can find pointers to Heaviside's own writings and those of others in the Notes.

I hope this book will leave readers feeling glad to have met Oliver Heaviside. His best friend described him as 'a first-rate oddity'. I agree – with a heavy emphasis on 'first-rate'.

Acknowledgements

Many people have played a part in the creation of this book. I am very lucky to have had generous help from Alan Heather. He is a great-great-grandson of Oliver's uncle, George Heaviside, and has provided personal memories – his mother described Oliver as 'an awkward old cuss but a brilliant man' – and a great deal of information about the family and about Oliver's life in Devon.

Everyone who is interested in Heaviside today owes a great debt to the Chelsea Publishing Company and others for keeping their editions of his collected *Electrical Papers* and his treatise *Electromagnetic Theory* in print over the years, thus giving direct and easy access to the great man's thoughts. Thanks are similarly due to the Institution of Engineering and Technology for maintaining their fine collection of Heaviside's notebooks, letters and papers, and I am especially grateful to archivists Anne Locker and Asha Marvin for their efficient and friendly help with research.

I am also indebted to fellow writers, especially to Bruce Hunt for his illuminating account of Heaviside's relationships with Hertz, Lodge and Fitzgerald in *The Maxwellians*, to Ido Yavetz for his scholarly insight into Heaviside's work in *From Obscurity to Enigma*, and most of all to Paul Nahin for his monumental biography *Oliver Heaviside: Sage in Solitude*.* Nahin's book is a treasure house of information about Heaviside and his contemporaries. Why, then, write another book? I hope that mine will complement Nahin's and others by presenting a compelling story and so bringing Oliver to a wider audience than he has so far enjoyed.

Nahin acknowledges his own debt to Professor Ben Gossick, who amassed a vast collection of material in preparation for a biography of Heaviside but died in 1977 before completing it. It is thanks to them both that I have been able to include some poignant extracts from Oliver's letters to Fitzgerald, and my obligation extends to the Royal Dublin Society Library, from whose Fitzgerald collection Gossick had photocopied the letters.

Profound thanks are similarly due to Ivor Catt for editing and publishing G.F.C. Searle's intimate recollections of his friend Oliver in *Oliver Heaviside, The Man*. Searle's little book shines a clear light on Oliver's character and is the source of many of the passages I have quoted from Heaviside's correspondence with Searle and Lodge, including his letters to Lodge on candidacy for the Royal Society and the death of Fitzgerald.

---

* Retitled for its second edition – see Bibliography.

I am grateful to Harold Allan and to Bill and Margaret Crouch, who have with great patience and perception read through drafts and suggested improvements, and to Ernie Freeman, Bruce Hunt (again) and Dominic Jordan for their encouragement and advice.

The task of describing the historical background to Heaviside's work succinctly has been simplified by John Wiley and Sons kindly allowing me to include in the narrative several short passages from my earlier book *The Man Who Changed Everything: The Life of James Clerk Maxwell.*

It is fitting that this book on Heaviside should be published by the IET, who (as the Society of Telegraph Engineers) threw him out for not paying his subscriptions but later (as the Institution of Electrical Engineers) awarded him honorary membership and the Faraday Medal. I would like to thank the editorial and production staff, especially my editors Lisa Reading and Jo Hughes for their expert help and guidance.

The final word of thanks goes to my wife Ann for putting up with Oliver as well as me for the past five years.

Chronology
Principal events in Heaviside's life

1850   Born at 55 King Street (now Plender Street), Camden Town, London, 18 May

1858   Contracted scarlet fever; the occurrence left him partially deaf for life
Started at the High Street School, St Pancras

1861   Changed schools, to Camden House School

1863   Family moved to 117 Camden Street, Camden Town

1865   Took College of Preceptors Exam, youngest of 538 candidates; came fifth

1866   Left school; began two years of private study at home

1868   Started work as a telegraph operator at Fredericia, Denmark

1870   Posted by his company to Newcastle on promotion to Chief Operator

1872   **Published his first paper, 'Comparing Electromotive Forces'**

1873   **Published paper, 'On Duplex Telegraphy', which caused his first brush with senior Post Office engineer William Preece**

1874   Resigned his post and returned to London to live with his parents

1875   Moved, with his parents, to 3 St Augustine's Road, Camden Town

1876   **Published paper 'On the Extra Current', which greatly extended William Thomson's theory of the transmission line**

1882   Began his long association with *The Electrician* when he was asked by the editor to contribute regular articles

       Published first paper in the series 'The Relations between Magnetic Force and Electric Current', in which he introduced vector analysis

1885   **Published first paper in the series 'Electromagnetic Induction and its Propagation', in which he reformulated Maxwell's theory of electromagnetism and introduced the four famous 'Maxwell's equations'**

       In the same series, gave the formula for energy flow in an electromagnetic field and showed that electrical power doesn't flow *in* a wire but in the space alongside it

1887   **Announced the formula for eliminating distortion from telephone lines in his series 'Electromagnetic Induction and its Propagation'. A paper giving a fuller treatment had been suppressed by William Preece**

       Published paper 'On Resistance and Conductance Operators', which transformed circuit theory with its generalisation of Ohm's law

1888   **Published series of papers 'On Electromagnetic Waves'**

Received the first public recognition of his work when it was praised and publicised by Oliver Lodge and George Francis Fitzgerald

1889   Received further recognition when Sir William Thomson used his Presidential Address to the Institution of Electrical Engineers to extol his work

Moved with his parents to 15 Palace Avenue, Paignton, Devon, to live in a flat above his brother Charles' music shop

1891   Elected Fellow of the Royal Society

1892   **Published *Electrical Papers,* a two-volume retrospective collection of his written work to date**

Published Volume I of his treatise *Electromagnetic Theory*

1893   **Published Parts I and II of his paper 'On Operators in Physical Mathematics', in which he described his controversial operational calculus, but the Royal Society rejected Part III the following year because it thought the mathematics insufficiently rigorous**

Proposed inserting inductive coils to approximate the distortionless condition in telephone lines in his paper 'Various ways, good and bad, of increasing the Inductance of Circuits'

1894   Refused an offer of financial help from the Royal Society

His mother, Rachel, died

1896   Granted a Civil List Pension of £120 a year

His father, Thomas, died

1897   Moved to Bradley View, Totnes Road, Newton Abbot, to live on his own

1899   **Published Volume II of *Electromagnetic Theory***

1900   US patent granted to Michael Pupin for a method of reducing distortion in telephone lines, based on Heaviside's formula of 1887 and his suggestion of 1893 for inserting inductive coils in lines. Pupin made a fortune; Heaviside got nothing

1902   **Predicted the existence of an ionised layer in the upper atmosphere which reflected radio waves; it became known as the *Heaviside layer* when the prediction was confirmed in the 1920s**

1904   Selected to receive the Royal Society's Hughes Medal, but refused it

1905   Awarded honorary doctorate by the University of Göttingen

1908   Moved to Homefield, Lower Warberry Road, Torquay, to live with Mary Way, sister of his brother Charles' wife Sarah

Awarded Honorary Membership of the Institution of Electrical Engineers

1912 **Published Volume III of *Electromagnetic Theory***

Short-listed for the Nobel Prize for Physics, along with Einstein and Planck, but the Prize went to Niels Gustaf Dalen

1914 Civil List Pension increased to £220 a year

1916 Mary Way left, leaving Heaviside on his own at Homefield

1918 Awarded Honorary Membership of the American Institute of Electrical Engineers

1921 Awarded the Faraday Medal by the Institution of Electrical Engineers

1925 Found unconscious at home in early January and moved to the Mount Stuart Nursing Home

Died four weeks later, on 3 February

Buried with his parents in Colley End Road Cemetery, Paignton

Note: The papers mentioned here form only a small part of Heaviside's published work. Almost all of it has been collected in the two volumes of *Electrical Papers* and the three volumes of *Electromagnetic Theory*, which together run to more than 2500 pages.

Cast of main characters
Relations, friends and adversaries

Brock, Henry: Police Constable in Torquay whose family befriended Oliver in his last years

Fitzgerald, George Francis: Professor of Natural and Experimental Philosophy at Trinity College, Dublin; friend and work associate

Heaviside, Rachel (née West): Oliver's mother

Heaviside, Thomas: Oliver's father; engraver

  Heaviside, Herbert: Oliver's eldest brother, estranged from the family; Post Office clerk

  Heaviside, Arthur: Oliver's second eldest brother and occasional collaborator; telegraph and telephone engineer

  Heaviside, Isabella (née Bell): Arthur's wife

    Their children (eldest first): **Lionel, Basil, Colin**

  Heaviside, Charles: Oliver's brother; music shop proprietor

  Heaviside, Sarah (née Way): Charles' wife

    Their children (eldest first): **Rachel, Ethel, Charles, Frederick, Beatrice**

Hertz, Heinrich: Professor of Experimental Physics at the Technische Hochschule, Karlsruhe; friend and work associate

Lodge, Oliver: Professor of Physics at University College, Liverpool; friend and work associate

Maxwell, James Clerk: Professor of Experimental Physics at Cambridge University who died in 1879 aged 48 and whose work inspired Heaviside, Fitzgerald, Hertz and Lodge

Perry, John: Professor of Mechanical Engineering at Finsbury Technical College; friend and work associate, strong supporter among the British electrical engineers

Preece, William: Engineer-in-Chief of the Post Office Telephone and Telegraph Department; Oliver was his bête noire and the feeling was reciprocated

Pupin, Michael: Professor of Engineering at Columbia University, New York; the man who gained glory, and money, from patenting a way of reducing distortion in telephone lines which was based on a formula of Heaviside's

Searle, George Frederick Charles: Lecturer in physics at Cambridge University; Oliver's closest friend

Tait, Peter Guthrie: Professor of Natural Philosophy at Edinburgh University, leading authority on quaternions; crossed swords with Oliver when he disparaged them

Thomson, William (later Baron Kelvin of Largs): Professor of Natural Philosophy at Glasgow University; patriarch of British physicists; gave Oliver support at several critical stages

Way, Mary: sister of Charles Heaviside's wife Sarah; owner of Homefield, Torquay, when Oliver came to live there with her

Wheatstone, Charles: Oliver's uncle by marriage; inventor and Professor of Experimental Physics at King's College, London; mentor to Oliver and his brothers when they were boys

Wheatstone, Emma (née West): sister of Oliver's mother Rachel; wife of Charles Wheatstone

Note: Most of the professional people listed here held various posts; the most relevant one is shown.

Introduction

Heaviside's whole life can be read as a kind of allegory on the ultimate triumph of virtue.

D.W. Jordan, *Annals of Science*

Communications engineering began with Gauss, Wheatstone and the first telegraphers. It received its first reasonably scientific treatment at the hands of Lord Kelvin ... and from the eighties on, it was perhaps Heaviside who did the most to bring it into a modern shape.

Norbert Wiener, *Cybernetics*

Oliver Heaviside, who lived from 1850 to 1925, was a true pioneer. Like his near-contemporaries in the American West he lived by his own code, operated in uncharted territory and never shirked a fight. His frontier was the nascent technology of electrical communications and his weapon was mathematics.

Heaviside's mathematics was like no other before or since. Scorning the formal rigour which runs in the veins of orthodox mathematicians, he found solutions of startling originality to real-life problems, such as how to make a distortion-free telephone line. He crafted James Clerk Maxwell's wonderful but hitherto inaccessible electromagnetic theory into a form which engineers and physicists could use in their everyday work and invented an 'operational calculus' by which complex differential equations could be solved as if by magic. He showed how to analyse any circuit and created the method of vectors that everyone now uses to analyse electromagnetic fields. In short, he founded much of the subject of electrical engineering as it is taught and practised today: every textbook and every college course bears his stamp.

Why, then, is his name not on everyone's lips? Probably because the ideas and methods he introduced are now taken for granted as though they had always existed – almost nobody wonders how they first came about. There is a poignant story here. Many of Heaviside's innovations are now part of the standard toolkit and no more contentious than the six times table but when they first appeared they were thought to be weird, even outrageous. They were so unlike anything that had gone before that many of the leading men of the time regarded them with horror; others were simply baffled. Heaviside battled long and hard against ignorance, prejudice and vested interests to get his ideas accepted. In the end he prevailed and brought about huge advances in electrical communications. More than this, he took the lead in creating what was, in effect, a whole new discipline – electrical engineering science – with its own language and its own way of looking at the physical world. But somehow, through the wayward channels of posterity, his name has been lost from view, and our great debt to him is rarely acknowledged today.

He was the youngest of four sons in a respectable but rather poor family in London, a bright but self-willed child who resisted his parents' exhortations to 'try to be like other people'. A severe bout of scarlet fever left him partially deaf – a condition that remained with him for life, as did a tendency to turn a deaf ear to unwelcome advice. He did well at school despite trying the patience of his teachers, never accepting anything just because it came from an established authority. After leaving school at sixteen he spent two years reading everything he could find on scientific topics before taking his first and only job – as a telegraph operator on the link between England and Denmark. This turned out to be a formative experience. While sending and receiving thousands of messages he got to know his equipment as a farmer knows his sheep. When things went wrong he diagnosed and repaired faults, sometimes spending a week at sea, grappling, cutting and splicing cable. In spirit he remained a telegrapher all his life – for him mathematics was simply an extension to the engineer's toolkit – but he began to see that the technology was in a rut and that the only way to make progress was to develop and bring in the new tools. Mathematically based research became his consuming passion and after six years he resigned his post, never again to take up regular employment.

From then on he lived what was, by normal reckoning, an extraordinarily dull life, rarely venturing further from home than his feet or his bicycle could take him. But his adventures were in the mind, and some of the ablest and most powerful people of the time became caught up in them. His letters and papers crackled with original and provocative ideas which stirred up many a stormy exchange of words in the journals. Oliver's strategy was always to pitch in boldly – tact was never in his repertoire – and he developed a style of literary invective to rival that of Samuel Johnson or Dorothy Parker. It was partly his own fault that his ideas were so slow to catch on – even mathematicians found his papers difficult and he rarely took the trouble to include step-by-step explanations to help the reader along. The few people who did manage to understand what Heaviside was up to were not engineers but academic physicists and they liked what they saw. After a while, self-taught Oliver was corresponding on level terms with some eminent professors. One was the great German physicist Heinrich Hertz, who performed one of the finest experiments of all time when he verified Maxwell's theory by producing and detecting electromagnetic waves. Sadly this friendship ended when Hertz died at the tragically early age of thirty-six.

But there were other friends. One was the Cambridge physicist G.F.C. Searle, who lived until 1954 and prided himself on being among the few people still alive to have met both Maxwell and Heaviside. Another was George Francis Fitzgerald, professor at Trinity College, Dublin, who was the first to predict that an object moving near the speed of light appears to contract along its direction of motion, and produced a formula which became part of Einstein's special theory of relativity. These friends were captivated by Heaviside's almost magical insight into the mysterious ways of electricity and entertained by his puckish sense of fun, but their feelings for him went deeper than this. Something in the make-up of their cantankerous colleague inspired unshakeable loyalty. Time after time he brusquely rebuffed their good advice – to make his papers easier to understand or to use a modicum of tact when he got into disputes – but none of this blunted their unwavering kindness and support. Oliver was lucky, too, in his dealings with *The Electrician*. This was a weekly magazine for engineers, almost none of whom could make head or tail of his articles. But a succession of far-sighted editors with saint-like patience published them regularly for many years and his collected papers are now classics, still in print.

Heaviside's earnings from fees and royalties were meagre and he was always poor. He was convinced that his work was worth 'a pot of money' and always hoped that, one way or another, his ship would come in, but it never did. His method for eliminating distortion from telephone lines did make a fortune, but not for him. Some friends went to a lot of trouble to persuade the Royal Society to offer him an honorarium, but when he found out that it was to come from a fund 'for the aid of Scientific men and their Families as may from time to time require assistance' he turned it down. The least whiff of charity was hateful to him. The same friends later managed to get the Government to offer him a pension of £120 per year from the Civil List. He didn't refuse this one; after all the Duke of Wellington had accepted *his* pension from a grateful country, so it was right to do the same.

He moved from London to Devonshire with his parents in 1889 and spent the rest of his life there. Friends visited when they could. One recalled bicycle rides when Oliver used to 'scorch' downhill through the narrow lanes on his fixed-wheel machine with feet up on the front forks while the pedals whirled round on their own. He was equally fearless, or foolhardy, in his dealings with the established authorities of the day and despite being nominated for a Nobel Prize, and receiving honours such as Fellowship of the Royal Society and an honorary doctorate from the University of Göttingen, he always thought of himself as an outsider.

He never married and had no children but it would not be unreasonable to call him the father of electrical communications engineering. He deserves to be better known, and not just for his amply demonstrated genius. Everybody loves an eccentric and everybody loves a small fellow with the spirit to fight the rich and powerful. Heaviside had both these characteristics in good measure and his faults serve only to etch his memory in the mind. He has the makings of a popular hero. I hope this book will give him a chance.

Chapter 1
Do try to be like other people
London 1850–68

The nineteenth century was halfway through its run when a fourth child was born to Thomas and Rachel Heaviside. Their three older children were all boys: Herbert, aged eight; Arthur, five; and Charles, three. The baby was another boy and they called him Oliver.

They lived at 55 King Street, Camden Town, about a mile to the north of Euston railway station in London.[1] Their house was in what would now be called the Georgian style, with three storeys and a basement, but it was a poor specimen. Two of the eight window spaces at the front of the house had been bricked in to avoid window tax and a pretentious portico over the door did little to improve the air of drabness. But their accommodation was palatial in comparison to that of families who lived above the local shops or in lodging houses – sometimes with only one room. And nearby was one of London's most squalid areas, occupied by people whom a near-contemporary sociologist described as 'Lowest class. Vicious, semi-criminal.'[2]

The Heavisides were middle class but their hold on this status must have seemed precarious at times. Thomas, who came from Stockton-on-Tees in north-east England, was a talented wood engraver whose services had been much sought after – he had engraved illustrations for *The Pickwick Papers* in *The Strand* magazine – but the growing popularity of photography had eroded demand for his work and commissions were hard to come by. To make things worse, he had poor health and sometimes had to ask his brothers, also engravers, to help him finish jobs he had taken on. Rachel was from Taunton, in the south-west, and had left home to work as a governess. Like many people in fast-growing London, Thomas and Rachel had abandoned their roots to make a life in the noisy, crowded city. It was an uneasy kind of life. For those with industry, skill and luck there was the possibility of earning a lot of money and climbing a rung or two up the social ladder. But there was also the risk of sinking into poverty, with no safety net to limit the fall and give breathing space for recovery. The dread of being committed to a debtors' prison or, worse still, to the workhouse was an ever-present backcloth to the daily round.

With her husband's erratic and diminishing income, Rachel had a constant battle keeping the wolf from their door and Oliver's arrival made things still harder. She decided to turn her experience as a governess to account by opening a day school for girls. It meant giving over the best part of the house to the business and buying desks, chairs and books with no certainty of recouping the cost, but boldness was rewarded: she found pupils, classes began, and money came in. This relieved the financial situation but a feeling of unease still pervaded the house. Thomas was a proud man and his naturally hot temper was kept on the boil by a sense of failure: he could not provide fully for his family and the school was a constant reminder of his inadequacy. The boys often caught the rough edge of his tongue and were beaten for minor transgressions. There was not much tenderness from their mother either as the worry of running both the school and the household took its toll on her spirits. The ethos in the Heaviside home seems to have been one of duty, but love must have been there too, even if it was tightly buttoned up; the boys were well clothed and nourished and had the best education their parents could afford. And perhaps there was some banter along with the scolding – the deliciously irreverent vein of humour that runs through Heaviside's writings cannot have come from nowhere.

London was rife with infectious diseases and young children were especially vulnerable. Many died from such illnesses as smallpox, scarlet fever, typhus and typhoid. It was scarlet fever that struck Oliver. He fought it off but it cast a shadow on his life by leaving him partially deaf. As we shall see, he never acquired what would now be called social skills and always tended to see himself as an outsider. This oddity may have stemmed from his early days when he tried to join in street games with the local children but found himself excluded because he couldn't hear what they were saying. He was thrown onto his own resources and began to build a defence against the vexations of life. A stubborn independence seemed to take hold of him, almost against his will. He later wrote – at the start of a book[3] on electromagnetic theory:

> The following story is true. There was a little boy, and his father said, 'Do try to be like other people. Don't frown.' And he tried and tried but could not. So his father beat him with a strap; and then he was eaten up by lions.

The rapacious 'lions' were the orthodox mathematicians who had rejected his work because it did not conform to their standards of rigour.

From his earliest days he never fitted into a mould: all efforts to persuade or force him to conform to accepted beliefs or practices were doomed. His teachers had a hard time of it. When Oliver was five he started attending his mother's school. At first he objected to being put among the girls but one visit to the rough, noisy school round the corner was enough to show him his parents were right this time. At eight they sent him to a proper boys' school – the High Street School, St Pancras – which had been founded to provide education for boys 'of the intermediate classes'. At eleven he moved again, to the Camden House School. This was a first-rate establishment which entered its pupils for the College of Preceptors examination, a forerunner of today's national exams in secondary education. Oliver studied thirteen subjects, including English, Latin, French, physics, chemistry, and mathematics. By the standards of the time this was good, mind-broadening stuff and he learned fast under a diligent teacher, Mr Cheshire, who managed the rare feat of gaining Oliver's whole-hearted respect. He failed, however, to change his pupil's opinion that some of the lessons were a waste of time. Those that roused Oliver's greatest contempt were on Euclidean geometry. Extraordinary! How could the owner of one of the finest mathematical brains of the age take so strongly against the standard way of teaching a basic part of his subject?

His explanation, written many years later, tells us a lot not only about his ideas on teaching but more generally about his whole approach to mathematics.[4]

> Euclid is the worst. It is shocking that young people should be addling their brains over mere logical subtleties, trying to understand the proof of one obvious fact in terms of something equally … obvious, and conceiving a profound dislike for mathematics, when they might be learning geometry, a most important fundamental subject … I hold the view that it is essentially an experimental science, like any other, and should be taught observationally, descriptively, and experimentally.

The lessons on grammar came in for similar scorn and his comments on them show that he was already developing a fierce contempt of pomposity in all its forms.[5]

> I always hated grammar. The teaching of grammar to children is a barbarous practice, and should be abolished. They should be taught to speak correctly by example, not by unutterably dull and stupid and inefficient rules. The science of grammar should come last, as a study for learned men who are inclined to verbal finicking. Our savage forefathers knew no grammar. But they made far better words than the learned grammarians. Nothing is more admirable than the simplicity of the old style of short words, as in A sad lad, A bad dog, of the spelling book. If you transform these into A lugubrious juvenile, A vicious canine, where is the improvement?

While his mother Rachel was working as a governess before her marriage, her elder sister Emma took a step that turned out to have a profound effect on the lives of Oliver and his brothers. She became cook to the famous scientist and inventor Charles Wheatstone. This was a congenial job but it turned to something else when she and her boss succumbed to mutual attraction and she became pregnant. They married and three months later had their first child. Wheatstone is remembered today chiefly as the supposed inventor of the Wheatstone bridge – a type of electrical circuit used to compare resistances. Curiously, this bridge was actually the creation of a fellow scientist, Samuel Christie, but Wheatstone was indeed a prodigious inventor as well as a spectacularly successful businessman. Starting in the family tradition as a maker of musical instruments, he took up an amazing variety of interests and turned most of them to good account by way of patents and business ventures. He also held the post of Professor of Experimental Physics at King's College, London for many years, even though he took little part in college life – presumably they valued the kudos of having him on the staff.

In the ordinary way, the young Heavisides would have had a rather limited view of life and its prospects. Boys from lower middle class families rarely had the knowledge or the confidence to take up occupations different from those of their fathers or uncles and, even if they did, opportunities were hard to come by without some form of patronage. Oliver and his brothers had the best available local schooling but their father and paternal uncles were in the dying trade of wood engraving and not in a position to give the boys much in the way of inspiration, guidance or help when it came to choosing a career and getting started. It was their Uncle Charles who opened up the world for them. It was not just a matter of being inspired by his glittering array of inventions, which included such diverse items as the stereoscope, the five-needle telegraph, the English concertina and the Playfair cipher;[6] he was worldly-wise and a welcome source of down-to-earth advice. What is more, he was able to use his influence to see that they had a good start in whatever field they chose.

The Wheatstones lived in a grand house on the southern edge of Regent's Park, about half an hour's walk from Camden Town. Visiting them must have been like entering a different world. But in 1863 the Heavisides came into a small legacy and were able to move a short distance to 117 Camden Street, a better house in quieter surroundings. By then Rachel had given up her school and had started instead to take in paying guests, a scheme which turned out to be less work, and more lucrative to boot.

Oliver was delighted at the move. Many years later he summed up his early life in a letter to a friend.[7]

> I was born and lived 13 years in a very mean street in London, with the beer shop and baker and grocer and coffee shop right opposite, and the ragged school just around the corner. Though born and raised in it, I never took to it, and was very miserable there, all the more so because I was so exceedingly deaf that I couldn't go and make friends with the boys and play about and enjoy myself. And I got to hate the way of tradespeople, having to fetch things, and seeing all their tricks. The sight of the boozing in the pub made me a teetotaller for life. And it was equally bad indoors. A naturally passionate man [his father], soured by disappointment, always whacking us, or so it seemed. Mother similarly soured by the worry of keeping a school. Well, at 13, some help came, and we moved to a private house in a private street. It was like heaven in comparison and I began to live at once.
>
> C. Dickens, when he was at the blacking manufactory, lived in a lodging house just around the corner and I know exactly how he got his intimate knowledge of the lower middle classes.

A grim tale, but Heaviside always tended to caricature aspects of his own life and this picture probably doesn't tell the whole truth. His parents cannot have been ogres, or he would not have returned home to live with them for twenty years after leaving his job. He had the companionship of his brothers and was not too deaf to share in the family enjoyment of music: one of his earliest memories was of going to the piano and picking out the tune of 'Pop Goes The Weasel' with one finger. He came to love Schubert and Beethoven and taught himself to play the piano after a fashion. True to form, he didn't think much of conventional musical notation and invented his own. Drawing was another favourite pastime. Two surviving drawings, made when he was about eleven, showed that he had inherited much of his father's talent and was already a sharp observer of the world around him. He also made childish experiments, using whatever he could find around the house. In one, he found he could make waves travel along the clothes line in the back yard by wiggling one end. Nothing remarkable so far, but he didn't stop there. He found that when he tied knots in the line, spaced fairly closely together, the waves travelled far more strongly and smoothly. This discovery cannot have cut much ice with his mother, who had enough to bear without such unnecessary vexations, but it turned out to be highly relevant to his later work. By putting knots in the rope he was, in effect, adding small concentrations of mass and this process is a close analogy to what became known as 'inductive loading' in telephone lines – inserting coils of high inductance at intervals along the line to improve the transmission of speech. It was this method that transformed communications by making long-distance telephony possible, and Heaviside was the man who first proposed it.

At fifteen, he took the College of Preceptors examination, the youngest of 538 candidates. He won the prize for the highest marks in the natural sciences and came fifth overall, despite scoring only 15 per cent in Euclidean geometry – presumably he just couldn't bring himself to write the answers in the formal style he hated. A splendid result, but the school could take him no further. His parents could not afford to send him to university, and the natural next step was to find a job. But Oliver had absorbed enough at school to show him how much more there was to learn and decided to educate himself further by studying at home. This idea cannot have pleased his parents at first and it is likely that Wheatstone, who had learnt his own science by self-study, pitched in on his behalf. The outcome was that Oliver embarked on a two-year programme of self-education – a kind of DIY sixth form course – with the idea of taking a job at the end of it, probably in the telegraph business. At Wheatstone's suggestion he learnt the Morse code and a little Danish and German, but for the most part he followed his own principles: explore, observe and experiment. He later described the process in another of his autobiographical parables.[8]

> More than a third of a century ago, in the library of an ancient town, a youth might have been seen tasting the sweets of knowledge to see how he liked them … In his father's house were not many books, so it was like a journey into strange lands to go a book-tasting. Some books were poison: theology and metaphysics in particular; they were shut up with a bang. But scientific works were better; there was some sense in seeking the laws of God by observation and experiment, and by reasoning founded hereon. Some very big books bearing stupendous names, such as Newton, Laplace, and so on, attracted his attention. On examination, he concluded that he could understand them if he tried.

By this time his three brothers had left home to make their way in the world. Herbert, the eldest, had walked out after a row with their father. We don't know what the quarrel was about but it was bad enough to cause a permanent rift – he never saw his parents again. The only hint we have of the nature of the dispute is that the other brothers took their parents' side. Poor Herbert was in the wilderness. Perhaps Uncle Charles Wheatstone helped him; he became a telegraph clerk in Newcastle upon Tyne, where he married and brought up a family.

Wheatstone certainly had a hand in launching the careers of the other three boys. Oliver's second brother Arthur had also gone to work in telegraphy, though in happier circumstances, and was also in Newcastle. Arthur was a talented telegraph engineer and, unlike Oliver, a team player. He made steady progress in his profession and eventually rose to one of the top jobs. He and Oliver stayed close and collaborated on some innovative schemes, but, as we shall see, Arthur sometimes had to take care not to support his young brother's controversial ideas too openly for fear of jeopardising his own career prospects.

Brother Charles was a gifted and enthusiastic musician, and had a head for commerce. Wheatstone's original business, making and selling musical instruments, was still flourishing and it was natural for Charles to start work there. It was not long before he struck out on his own, taking a post as assistant in a music shop in Torquay, Devon. The venture prospered. He was soon a junior partner in the firm and by the time he became senior partner the business was doing well enough to open a second shop in nearby Paignton. This shop, or rather the flat over it, eventually became the home of Oliver and his parents.

By 1868, eighteen-year-old Oliver was ready to face the world, fortified by his two years of exploratory 'book-tasting'. A brief spell as assistant to his brother Arthur in Newcastle gave him a gentle introduction to the telegraph business but he soon spread his wings. With his good exam results – and, no doubt, Wheatstone's recommendation – he was offered a post as a telegraph operator for the newly formed Danish–Norwegian–English Telegraph Company at the excellent starting salary of £150 per year. It was an exciting time to join the company as their Anglo-Danish undersea cable was about to be opened. He accepted, packed his trunk, and set off for Denmark.

Chapter 2
Seventy words a minute
Fredericia 1868–70

Oliver's destination was the town of Fredericia on the eastern side of Jutland, where the company had set up its headquarters. There he joined a small group of English staff who worked alongside their Danish colleagues. The newly laid 420-mile-long North Sea cable ran from Newbiggin-by-the-Sea to Sondervig and was connected by overland lines to the main operating stations in Newcastle and Fredericia. The Anglo-Danish link was just a small part of a rapidly growing web of telegraph lines that already covered much of the world.

It was a very different world from that of 30 years earlier, when a few ingenious and enterprising pioneers launched the first commercial telegraph projects. News which would have taken many days to travel by land and sea now took only a few hours to arrive. People could read in the newspapers about yesterday's events rather than those of a week or more ago, and the conduct of national and international affairs, whether military, diplomatic or commercial, was transformed. One instance is enough to demonstrate this. Once its forces had gained the upper hand in the Indian Mutiny of 1857 and 1858, the British Government was able to send a message across the Atlantic in time to prevent two regiments from Canada embarking on an expensive and unnecessary journey.

Sub-ocean telegraphers were the technological élite of the day: cable testing rooms were, in effect, superbly equipped laboratories and much of the work, while serving a practical purpose, was, in fact, front-line scientific research. Heaviside's career as a practical telegrapher lasted only six years but it set the course of his life. In a striking parallel with Rudyard Kipling, whose seven-year spell in India as a young man provided the raw material for a lifetime's work, Oliver was enthralled by what he saw and heard. He felt compelled to discover more and to write about his discoveries. In short, he found his muse. To understand the grand obsession that came to take over his life, we need to take a look at the 30-year history of the telegraph. Indeed, we should go back a little further to two key events that made the telegraph possible.

The first was in 1799, when the Italian, Count Alessandro Volta, invented the voltaic pile, or battery, which provided a source of continuous electric current: previously it had only been possible to store electricity in such devices as the Leyden jar, which released all its charge in one burst. Volta had not set out to produce currents: he merely wanted to show that his friend Luigi Galvani was wrong. Galvani believed that the electricity by which he made dead frogs' legs twitch came from animal tissue, but Volta thought it was generated by chemical action between different metals in the circuit. His first pile, or battery, built from repeated layers of silver, damp pasteboard and zinc, was intended simply to prove he was right. It did, indeed, prove the point but the battery soon took on a life of its own and people started to use currents for such things as electroplating. Curiously, the name they gave to the phenomenon of continuous electric currents was not 'voltism' but 'galvanism'.

The second key event was in 1820, when the Danish physicist Hans Christian Oersted was lecturing to a class about electric circuits. He had left a compass on the bench near one of the wires and was amazed to see the needle jerk to a new position when he switched on the current. This was the first demonstration that electricity and magnetism are inseparably linked: every electric current wraps itself with an encircling magnetic field. By the same token, it showed how an electric current in a wire could be instantly detected: what you actually detect is not the current itself but the mechanical force exerted by its magnetic field on a nearby magnet – or, equivalently, on a separate current-bearing wire with its own magnetic field. Moreover, the strength of the force varied directly with the strength of the current. Will-o'-the-wisp electricity, which nobody really understood,[1] could now be made to produce the kinds of real forces with which everybody was familiar. The field was now open to ingenious inventors; they could use the mechanical force to move a needle, flip a switch, vibrate a diaphragm, or to work any other mechanism they could devise which would produce a recognisable response.

The two main ingredients for the electromagnetic telegraph were there: a battery that could be used to generate a current, and the knowledge that the current could be made to operate a mechanical device. But more was needed – the imagination to see that electricity could carry messages, the ingenuity and perseverance to produce a working system at reasonable cost, and the enterprise to harness the commercial potential. The people who came to the fore were, for the most part, not trained scientists. In the early days of the telegraph nobody knew much about the physics of electrical communications, so there was no particular advantage in being an expert. Flair and determination were the cardinal qualities, and luck. In the mid-1820s the Russian, Baron Pavel von Schilling, devised what was possibly the first practical scheme.[2] After failing to rouse his government's interest he spent ten years travelling round Europe demonstrating his signalling apparatus. It was a popular road show but nobody took up his ideas. At length the Russian government asked him to set up a telegraph link from St Petersburg across the Gulf of Finland to Kronshtadt, but he died within two months of getting the commission.

The unfortunate Schilling had unknowingly played a part in getting the telegraph under way in Britain. In 1836, Professor Muncke of Heidelberg University gave a lecture on electricity which included a demonstration of Schilling's system. In the audience was a 34-year-old former officer in the East India Company's army, William Fothergill Cooke. He was in Heidelberg to study, of all things, anatomical modelling in wax – a technique in demand in India, where native-born trainee surgeons had religious objections to practising on corpses. Cooke knew little of electricity and may have gone to Muncke's talk simply for entertainment but it changed his life – he saw at once the huge potential of sending information long distances over wires and set to work trying to make a suitable device for receiving signals.

Cooke's early results with rough models were so good he thought he had found the entrance to a gold mine and became obsessed with keeping it secret. He returned to London to get the equipment professionally made but told his mother not to let anyone know what he was up to. He even had different parts of the mechanism made by different mechanics. The work proved troublesome and expensive but he eventually put together a prototype system and asked Michael Faraday at the Royal Institution for an opinion. Faraday said the design seemed sound, but could give no answer to the crucial question of distance – how long could the wire be before the signal became too weak to operate the receiving device? After an unsuccessful request to be allowed to run a trial on the Manchester and Liverpool Railway, Cooke learned that he had a formidable rival, Charles Wheatstone. Jaded and short of money, he began to think that cooperation might be more productive than competition. It turned out that Wheatstone was of the same mind; they decided to pool their ideas and tested Cooke's apparatus over 4$1/2$ miles of wire in Wheatstone's laboratory. It didn't work.

They turned their attention to the receiving device. The problem with Cooke's receiver was that it needed more current than the battery could supply when a signal was sent over a long line. The most sensitive kind of receiver was the simple galvanometer, in which the current's magnetic field was used to deflect a compass needle – essentially Oersted's original method, but much refined by Ampère and others. Cooke's first idea had been to employ galvanometers, but after they proved rather delicate and awkward to use he had concentrated on mechanisms worked by electromagnets that had a soft iron core inside a wire coil and operated independently of the earth's magnetic field. One such was the clockwork alarm – simply an alarm clock that rang briefly each time a pulse of current in the line activated an electro-magnetic switch. Receivers based on electromagnets were robust and versatile and would come to be the generally preferred type, but these were early days and for the moment the partners decided to ditch Cooke's system in favour of a galvanometer scheme that Wheatstone had been working on, the five-needle telegraph. They took out their first joint patent in 1837 and caught the ear of Robert Stephenson, Engineer to the London and Birmingham Railway, who sponsored a trial on Camden Bank, a sharp incline out of Euston Station up which trains were pulled by mechanical cable. Stephenson was impressed and tried to persuade his directors to extend the telegraph all the way to Birmingham, but they fought shy of the new-fangled wizardry and did not even renew the Camden project when the six-month trial ended.

But as that door shut another opened. Isambard Kingdom Brunel was building the Great Western Railway and commissioned Cooke to set up a pilot telegraph system along 15 miles of track from its London terminal, Paddington, to West Drayton. It worked so well that Cooke didn't wait for the GWR's decision but extended the telegraph a further 6 miles to Slough at his own expense. Fortune favoured the brave: Cooke's enterprise was soon rewarded with some priceless publicity. John Tawell, who had committed a murder in Slough, had been spotted making his escape by train but was caught by police at Paddington thanks to a telegraph message. An important part of the message was that Tawell had disguised himself as a Quaker. This posed a problem because the telegraph system was then using a device of Wheatstone's which transmitted only 20 letters of the alphabet, two of the missing letters being Q and U. So the Paddington police had to interpret the strange word KWAKER – fortunately they worked it out in time to arrest Tawell.

People liked nothing better than to read in the newspapers about murderers being caught, and the story fired their imagination. The telegraph was now firmly embedded in the public mind as a good thing. Government and commercial interest was growing, too, and the partners forged ahead. They had, by now, patented several kinds of instrument for sending and receiving messages, and Cooke had found that galvanised iron wires strung on wooden poles made cheap and serviceable paths for the signals. They opened a line between London and Portsmouth, half-funded by the Admiralty, and got a contract for another between London and Dover. But it was an uneasy partnership as they both claimed priority for some of the inventions. Their first quarrel was patched up when an arbitration panel split the honours evenly, but when they fell out again it was time to part. In 1845 Cooke formed the Electric Telegraph Company, with himself as one of the directors, and the company bought the rights to their joint patents. By 1848 it had 1800 miles of line. For a while capital spending ran so far ahead of receipts that the directors were in trouble, but an outstanding new engineer, Edwin Clark, got them back on an even keel. By 1851 the company was making £50,000 per year and the mileage had grown to 11,000.

Other countries were keeping pace. By now overland lines spanned Europe and were beginning to extend eastwards to Russia and to the Middle and Far East. The telegraph had burgeoned in America, too. The prime mover here was Samuel Morse, and it was he who brought about a huge improvement in signalling with his brilliantly simple Morse code. A sequence of mixed long and short pulses, dots and dashes, could now be used to send any text message. It was a good code – relatively easy to learn yet a source of satisfaction and pride to those who mastered it. Most of all it gave telegraphy a boost by establishing a common currency across countries and brought into sharp focus the great challenge now facing telegraph engineers: how to link countries by sending messages across the sea.

It was a relatively simple matter to send signals over land. Just string an iron wire on wooden poles to connect the transmitting and receiving stations and connect a battery. You could even use the earth itself to provide a return path for the current. The live wire needed to be well insulated but air was a splendid insulator, at least when dry, and you could stop current leaking to ground through rain-soaked wooden poles by fixing the wire on porcelain or glass mounts. Insulation in water was a much harder proposition: the wire on the seabed needed to be wrapped in a substance that could be easily worked into shape but would not crack, warp, perish, or lose its insulating properties over time. Wheatstone and other experimenters tried everything they could think of. One method was to embed the wire in a rope soaked in boiled tar; another was to wrap it with cotton inside a rubber tube inside an outer sheath of lead or iron. Sometimes the telegraph link worked well for a while but it always broke down. Then, in 1848, someone sent some samples of a new material to Michael Faraday, who reported enthusiastically to the *Philosophical Magazine* on its qualities – an excellent insulator, water-resistant, flexible, resilient and easily moulded to shape when warm. It was gutta percha, a tree gum from Malaya, and everyone wanted it, especially once the German engineer Werner Siemens had found a way to mould the material around a metal wire in a continuous run.

Gutta percha made trans-ocean telegraphy possible but laying cables on the seabed was an entirely new kind of venture, difficult and risky. It was hard to keep a rolling, pitching ship on course while dropping cable off the stern, especially when it was heavy-laden with great reels of cable. Going too far off-course led to the expensive humiliation of running out of cable, as several early expeditions found out. And paying out the cable was itself a tricky operation even in the calmest sea: let the cable drop too fast and you waste it; apply too much brake and you snap it. Early adventurers learnt such things the hard way and the first two working cables laid across the English Channel in the early 1850s, from Dover to Calais and Ostend, owed much to the sheer determination of the project leader John Brett, a former antiques dealer with no engineering training. More cross-Channel links soon followed and the Irish Sea, too, was crossed by cable.

The next challenge was both obvious and daunting: to lay a cable across the Atlantic Ocean. The intrepid Englishman Frederic Gisborne had embarked on a project to lay a chain of overland and underwater cables from the eastern United States to the tip of Newfoundland, from where it was only 2000 miles to the west coast of Ireland. Gisborne almost bankrupted himself in the process but his vision of an Atlantic cable was picked up by the New York paper magnate and entrepreneurial genius Cyrus W. Field, who formed the Atlantic Telegraph Company in 1856, with John Brett as President, and managed to sell the total share issue of 350 shares at £1000 each within two weeks. One of the subscribers was William Makepeace Thackeray. For the job of chief electrician the company chose the splendidly named Edward Orange Wildman Whitehouse, a doctor turned telegrapher from Brighton, who was energetic, ingenious and resourceful but dangerously ignorant on scientific matters. Wisely, they decided also to engage a scientific consultant, William Thomson of Glasgow University.[3]

Thomson was a star. Having gained his professorship at 22 and Fellowship of the Royal Society at 26, he was already, at 32, an immensely respected and influential figure. Everything seemed to come easily to him, from mathematics to technological design, and he went on to become, as Lord Kelvin, the patriarch of British science. A man of great charm and infectious enthusiasm, he inspired and helped many younger people, including Maxwell and, later, Heaviside.

Like all good scientists, Thomson had a burning curiosity about the physical world and when he heard of a puzzling new phenomenon in undersea telegraph cables he quickly set to work on an explanatory theory. Within a few days he had his theory and it held sway for 20 years; as we shall see, it was Heaviside who then moved things on by showing that Thomson had painted only part of the picture. The problem telegraphers had found was that signals lost their sharpness when sent through underwater cables: they became smeared out, so that the trailing part of one pulse overlapped the leading part of the next one. To keep successive pulses distinguishable, the time interval between them had to be increased, thus slowing the rate at which messages could be sent. And the longer the cable, the worse the effect. At first, telegraphers had no idea why this happened; then Michael Faraday put forward the idea that the cable acted as a giant electrical store, or capacitor, which took time to charge and discharge.

His brilliant insight was that while the cable was charging or discharging most of the action took place not in the central copper wire that sent the current, nor in the outer iron sheath that returned it, but in the insulating material between them. The material reacted to an electric force like a spring reacts to a mechanical force, building up strain and storing energy. When the battery was connected by depressing the sending key, the initial flow of energy went into 'stretching the spring' in the insulating material rather than driving a current along the wire. The current began to flow to the receiver as soon as the insulating material took some of the strain but was small at first and did not reach its full value until all the strain had been taken up. And when the sending key was released the current did not stop immediately but continued to flow until all the stored-up energy in the insulating material had been released. A rough analogy is pulling a sledge through snow with an elastic tow rope. As you start pulling, the rope stretches, absorbing and storing energy; the sledge starts to move as the rope begins to take the strain but doesn't reach full speed until the rope is fully stretched. And when you stop pulling, the tensile force in the stretched rope continues to draw the sledge along until the rope is back to its normal length.

Faraday was universally acknowledged to be a superb experimenter but many scientists thought his theoretical ideas on electricity vague and fanciful because he knew no mathematics and could not quantify the effects he described. Thomson, however, thought that Faraday's theory of the cable rang true and set out to give it mathematical expression. Drawing on a remarkable analogy he had used before, he derived an equation for the transmission of an electrical pulse along an underwater cable that had the same mathematical form as the one for the diffusion of heat along a metal bar. A great breakthrough, but it brought bad news for the Atlantic telegraph project. Thomson had found that the signal retardation – the amount by which the cable slowed the transmission of messages – didn't just increase with cable length, it was proportional to the square of the length. Other factors being equal, signalling speed on a 2000-mile cable would be 25 times slower than that on a 400-mile cable.

This unwelcome result came about because the retardation depended on the product of the cable's resistance and its capacitance, each of which was itself proportional to the length. There was already the worry that signals over 2000 miles of cable might be too weak to detect and now, it seemed, the signalling rate might be too slow to be profitable. But Whitehouse, the 'practical man', set no store by mathematics. He airily dismissed Thomson's theory, claiming that he had done tests using 1000 miles of underground cable which showed that signalling speed varied directly with cable length rather than its square. The two egos clashed. Thomson criticised Whitehouse's rough-and-ready experimenting and held to his own views, but had to defer to the more experienced man on matters of practical telegraphy. They declared an uneasy truce as Field had already bought 2500 miles of cable and the project was under way.

Things did not go smoothly. After two gallant but fruitless voyages the company's ships *Agamemnon* and *Niagara* had succeeded only in scattering several hundred miles of expensive cable on the ocean floor. At this point many of the project's backers were ready to cut their losses but the admirable Field steadied their nerves and prepared for another attempt. Soon the two ships were back in mid-Atlantic, each laden with enough cable for 1000 miles. After splicing their cable ends *Agamemnon* sailed for Valentia on the west coast of Ireland and *Niagara* for Trinity Bay, Newfoundland. Both reached their destinations on 5 August 1858 and the first Atlantic cable link was complete. After a worrying six weeks of testing and adjustments the cable was opened for use and messages flowed. Jubilation! New York held a great parade and a banquet for Cyrus Field. Hats were in the air but no sooner had they landed than the message traffic ceased. The company's engineers desperately searched for possible faults in parts of the cable near the shore stations and fobbed enquirers off with optimistic reports but it was soon plain to all that the Atlantic link was dead.

Faced with this huge disappointment and the inevitable plunge in share price, the directors gritted their teeth and set about trying to find out what had gone wrong, and what, if anything, was salvageable. It turned out that the cable had never worked well; not only was the signalling slow, messages had often had to be repeated several times before they made sense. Field took a close look at some cut-up sections he had sold as souvenirs to Tiffany's in New York and saw that the cable had been badly made – in places the copper wire was way off-centre, almost piercing the gutta percha insulation. And later investigation by Thomson found that the copper used was of poor quality, containing many impurities. Even with these flaws, the cable would have worked for longer had Whitehouse not wrecked it by using ruinously high voltages. As a later writer put it, 'he sent a stroke of lightning through the cable which required only a spark'.[4] Whitehouse was sacked but the damage had been done. Thomson at first defended Whitehouse: much as he deplored his colleague's methods he thought the man was following a genuine, if mistaken, belief in their superiority. But further investigation showed that Whitehouse had falsified the telegraph log, claiming to have received messages using his own equipment when he had, in fact, used Thomson's ultra-sensitive mirror galvanometer. He was not only a quack but a rogue, desperate not to be upstaged by a mere academic.

The project had taken a terrible knock: even the irrepressible Field could not kindle enthusiasm for another go. Then events way beyond the company's control took over; the southern states of America decided to secede and the country fell into a hideous civil war. The war put a stop to any thought of transatlantic cable-laying but engineers put the time to good use. They developed much better machinery for paying out cable from a ship's stern, and for picking it up again if needed. To lay the cable smoothly they needed an accurate chart of the mountains and chasms on the ocean bed and Thomson developed a sounding machine for the purpose. The Atlantic adventurers learnt, too, from successful projects elsewhere: the cable firm Glass, Elliot had successfully laid an undersea link of about 1500 miles from Malta to Alexandria and another across the Persian Gulf. At Thomson's instigation, manu-facturers improved cable design and brought in proper quality control to ensure, for example, that the electrical resistance was acceptably low. Meanwhile, Cyrus Field had kept the vision of an Atlantic cable alive in America and as soon as the civil war ended he began raising funds for another venture.

The drama took an unexpected turn when a new principal character appeared. Daniel Gooch, an English railway engineer turned tycoon, bought a five-year-old ship at a bargain price. She was Isambard Kingdom Brunel's *Great Eastern*, five times the size of any other ship then afloat and capable of steaming from England to Australia without re-coaling. Brunel had intended her to carry passengers around the world in style but she seemed to be ill-starred. At the ceremonial launch she refused to budge more than a few feet and became stuck on the ramp. It was months before they managed to get her into the water and when at last she steamed into the English Channel there was an explosion that killed several people and blew up the front funnel. Brunel himself died a few days later, his health shot to pieces by the stress of the venture. The *Great Eastern* plied the Atlantic for several years but failed to attract enough passengers or cargo to be profitable. Then, just as things were looking up, she broke her rudder in a storm and limped to Cork. The owners got her back into commission but repairs took eight months and cost £60,000. It was not long before fate struck again: she scraped a rock and suffered a great gash. Her double hull kept her afloat and she managed to get to Long Island Sound but this time the cost of repairs bankrupted the owners. Gooch saw his opportunity and raised the money to buy the ship for one-tenth of the sum it had taken to build her.

Gooch was a director of the Telegraph Construction and Maintenance Company, newly formed by a merger of Glass, Elliot with the Gutta Percha Company. He had seen that the *Great Eastern*, once suitably fitted out, would be an ideal cable ship and with his audacious purchase he had wrested the initiative from Cyrus Field. He offered to lay a transatlantic cable for Field's Atlantic Telegraph Company in return for £50,000 in that company's shares; his company would bear all the laying costs and hand over the cable on completion. Perhaps it was a relief for Field and the ATC to have someone else make the running for once. At any rate, they accepted.

In July 1865 the *Great Eastern* set sail from Ireland with 5000 tons of cable. William Thomson and Field's electrician C.F. Varley were on board but only as observers; they were under company orders not to give any advice or opinion except in a written reply to a written question. Even then they were to include a statement absolving their company from any responsibility for the consequences. The expedition was soon in trouble: a mysterious fault appeared in the cable as it was lowered into the sea. To everyone's relief this was soon fixed and the great ship showed her mettle, steaming serenely through heavy seas while the cable unreeled smoothly behind her. But Providence rarely smiled on the *Great Eastern* for long. A few days later another fault appeared and the cause turned out to be similar to the first – a sliver of iron had pierced the cable sheath. Sabotage? Some of the crew thought so but they got on with the job and unpleasant suspicions were dispelled when an alert crewman saw a splinter break off the cable and become lodged in the paying-out machinery. This was almost certainly the cause of the faults. He removed the splinter before it could do damage and for a while everything went smoothly. They were two-thirds of the way to Newfoundland when the dreaded words 'dead earth' were heard – another fault. The wearisome corrective procedure was to stop, turn the ship around, and then gently reel in cable over the bows until the fault was found. This was all routine by now but a sudden change of wind put the cable under extra stress and when the ship met a freak wave, not big but from an awkward direction, the cable snapped. The crew tried desperately to grapple and recover the cable but it was more than two miles down in one of the deepest parts of the Atlantic. Three times they managed to hook the cable but each time the grappling line broke before they could pull it to the surface. By then they did not have enough line to try again – the expedition was over.

Yet again, Field was the man to rally spirits, and funds. The next attempt would be sure to succeed as long as they had more and stronger grappling lines to pick up any loose cable. In fact, they did not need them. The *Great Eastern*, for once, made a blissfully uneventful voyage and laid a cable from Valentia to Trinity Bay in July 1866. This time there were no wild celebrations – just the satisfaction, and relief, of a solid success. It seemed the time to ride their luck, so they set out to recover and complete the 1865 cable. Thomson and the engineers had worked out a safer technique, using three ships to lift the great weight of cable. It worked. Now there were two Atlantic cables; both were soon busy with business and private telegrams, and investors in the Atlantic Telegraph Company at last had a return on their outlay.

This brings us to the time of Oliver's move to Denmark, but before picking up the thread of our main story we can look ahead a little to complete the tale of the *Great Eastern*. She was recommissioned as a passenger ship and again failed to attract passengers but her great days were not over. Chartered by the French government in 1869, she returned to what she did best and laid a third Atlantic cable, this time from Brest to Newfoundland. She followed up with two more Atlantic cables in the early 1870s but could not compete with the new, more economical ships that were specially built for cable work and she lay rusting in Milford Haven before being bought to act as a tourist attraction and floating advertisement hoarding in Liverpool. When she was scrapped in 1888 a story quickly took root that the source of her misfortunes had been found – the skeleton of a shipyard worker who had been trapped between the two layers of the double hull.

The sense of adventure which had inspired and sustained the early heroes of the telegraph was still strong in the air at the time Oliver arrived in Fredericia. Men like Cooke, Wheatstone, Morse and Field were international celebrities. Ordinary people began to see that the telegraph was for them as well as for the swells in government and big business, and kept the offices busy with telegrams. Even so, the practice of sending messages down a wire still seemed to them like a benign kind of witchcraft. Indeed, a cable testing room would have seemed to the untutored eye like a wizard's lair, stocked with the strange tools of the trade: rheostats, bridges, shunts, condensers and various mysterious devices for sending and receiving signals. Bright young men were drawn to the work by good wages and the excitement of being at the forefront of technology. They learnt their craft on the job: there was no coherent body of knowledge called electrical circuit theory, still less books or courses on the subject. Oliver had an advantage over his fellows: he would have seen similar equipment in Wheatstone's laboratory and had some idea what the various pieces of apparatus were for. The telegraphers were free to improvise and experiment with the apparatus; indeed they had to, simply to keep the lines working, and this was the way that Oliver learnt best. He quickly got to grips with the task and became an effective troubleshooter. But for him it was not just a matter of fixing the problem in hand. He became fascinated by the curious ways of electricity, which often baffled even the most experienced of his colleagues, and each of his investigations was a step in what became a grand quest for knowledge.

Like the other operators, Oliver spent most of his working day in the telegraph office, sending and receiving messages. They were kept busy. As he put it, 'the speed of working was always pushed to the greatest possible (the press of business being such as to make 25 hours work per day all too short)'. His deafness was not a handicap because the receiving devices used visual cues rather than sounds. The standard method employed paper tape, in which holes were punched to indicate the dots and dashes of Morse code. Wheatstone had invented a machine that read the tape and automatically sent the appropriate pulses down the line. At the receiving end a complementary machine converted the pulses back to marks on paper tape that the clerks could read. On landlines the Wheatstone automatic system could transmit messages at more than a hundred words per minute but over the North Sea cable the speed came down to thirty or so at best. Any faster and the receiving machine could not pick out individual pulses because they had become smeared-out in their passage along the cable. Oliver found that he and some of the other operators could do better if they used a different type of receiver – the siphon recorder. This was a wonderfully sensitive instrument invented for the Atlantic telegraph by William Thomson.* With the siphon recorder a rate of seventy words per minute was possible, but only with expert interpretation. At that speed the dashes in the pen-trace appeared as rounded humps, and the dots were mostly wiped out – to make sense of the message the operator needed to spot where they ought to have been!

The work of telegraph offices, especially those using undersea cables, had not yet settled into a smooth routine. With the imperative of keeping message traffic flowing, the operators had to become adept at making adjustments and fixing minor problems. The humdrum business of sending and receiving messages was, perforce, mixed with scientific research. Oliver and his colleagues got to know the idiosyncrasies of electricity but were often at a loss to explain them.

Two things in particular defied explanation. One was that the rate of signalling over the Anglo-Danish link was higher eastwards than westwards: messages could be sent from Newcastle to Fredericia 40 per cent faster than in the other direction. Nobody knew why. It seemed a reasonable guess that it had something to do with the landline being shorter on the English side but this turned out to be only part of the story. Deeper thinking was required and Heaviside supplied it. For the present he was as perplexed as anyone but once he had begun to think about a problem he never let go and, as we shall see, he found a complete and surprising solution a few years later.

Equally puzzling was that a cable fault sometimes *improved* the quality of the received signals. At some point in the cable the gutta percha insulation would become weakened, allowing current to leak between the central copper wire and the iron sheath. A fault had to be found and repaired as quickly as possible because it led to cable failure: eventually so much current leaked away that no signal at all got to the receiving end. But before the cable failed, the signals, though weak, would be clearer than on the fault-free line and could be sent at higher speed: a partial fault

---

* Thomson's original receiving device for the Atlantic telegraph, the mirror galvanometer, was even more sensitive than the siphon recorder but less practical because it left no record of the coded signal.

seemed to reduce the cable's tendency to turn sharp pulses into smeared-out blobs. These incidents were fleeting and dwarfed by the urgency of fixing the fault, so few people gave serious thought to the implications. Heaviside was one who did, and his thoughts eventually led him to a discovery that made national and international telephone networks possible: how to make a distortion-free line.

The members of the small band of English operators in Fredericia were thrown closely together. We know little about Oliver's relations with his fellows; he was never an outgoing person and it seems that he made no lasting friendships while in Denmark. But the camaraderie ran deep: he had started to forge an intense and lifelong feeling of comradeship with telegraphers and, more generally, with all who strove to put the strange properties of electricity to practical use. As to how he got on with his Danish hosts, we have a small clue from a letter he sent 42 years later to a friend who had been taken ill while on a Scandinavian holiday. Oliver wrote: 'Hvorledes hat de det nu? Jeg taenken du ei skulde drikket daarlig Vandet.' (How are you now? I think you must have been drinking bad water.) Not much, but enough to suggest that he had learnt to make basic conversation in Danish.

Oliver would certainly have met the company boss. C.F. Tietgen was an ambitious financier whose many commercial interests included sugar, beer and banking, but the telegraph was his favourite project. Tietgen's grand vision was the *Great Northern Chain*, in which Denmark would be the nerve centre of a huge telegraph system linking Britain and Europe with Russia and the Far East. The speed with which he accomplished this feat is astonishing. The Danish–Norwegian–English Company was one of three small companies he had set up in 1868 to lay undersea cables at the European end and get business going. The following year he merged them to form the Great Northern Telegraph Company and extended the operations eastwards by obtaining rights from the Tsar of Russia to lay and operate submarine cables anywhere on Russian territory, at the same time persuading the Russians to build a landline all the way from St Petersburg to Vladivostok. While work on this was under way the company was busy organising the Far Eastern end of its network, laying cables connecting China to Hong Kong and Japan. By 1872 Tietgen's vision was realised and his company's lines were buzzing with commercial traffic between Europe and the Far East. The Great Northern became one of the leading companies in Denmark and is still trading today.

After the 1869 merger, work in the telegraph office in Fredericia went on much as before except that the English operators began to be replaced by Danes. Perhaps it was company policy that brought about the change; perhaps it was simply that the Englishmen wanted to go home. Whatever the reason, Oliver soon found himself the sole survivor. What now? He had done a good job for his employers and they decided to post him to Newcastle where he would be promoted to chief operator and given a pay rise to £175 per year. This was a fine salary for a twenty year-old, almost twice that of his eldest brother Herbert who was working as a telegraph clerk for the Post Office. In January 1870 he left for Newcastle, having made what seemed to be a highly promising start to a career in telegraphy. Great things did indeed lie ahead, but along a path that no one could have predicted.

Chapter 3
Waiting for *Caroline*
Newcastle 1870–74

Oliver's reputation as a clever worker had probably preceded him to Newcastle, coupled with the knowledge that he was Wheatstone's nephew. First impressions were favourable: one of his new colleagues remembered him as 'a very gentlemanly looking young man, always well dressed, of slim build, fair hair and ruddy complexion'. He was quiet and often wrapped in his own thoughts but by no means insular: he was always ready to explain the workings of the apparatus to newcomers and to discuss technical matters with anyone who had a genuine interest. Writers about Heaviside have generally commented that he didn't make much of a mark during his four years with the Great Northern in Newcastle. This is broadly true, but he did his duty well and, as we shall see, saved the company money with some ingenious trouble-shooting. Looking at things the other way, these four years of telegraph work certainly made a mark on him.

With one exception in later life – and we'll come to that – Oliver always took the easy option when it came to domestic arrangements, so it was natural for him to lodge with brother Arthur and his wife Isabella. Arthur had joined the Universal Private Telegraph Company at Newcastle nine years earlier and when control of all inland telegraph services in England passed to the Post Office in 1870 he stayed on as a District Superintendent Engineer for his new employers. The two brothers got on well. In caricature, Oliver was the brilliant but wayward youngster and Arthur the steady, protective elder brother – a pattern that continued throughout their lives. In such congenial surroundings Oliver was able to begin making electrical experiments in his own time, following his intuition. Arthur was himself an inventive engineer and joined in when he could, but family duties had first call on his time – Isabella gave birth to their first two children while Oliver was staying with them, and was expecting a third at the time he left.

We can get some idea of what Oliver was up to from a letter he wrote half a century later to the parents of Gordon Brown, a young physicist who was killed in the First World War. They had sent Oliver a copy of a paper of Gordon's and he replied: 'For a youth of 20 he was surprisingly advanced. Why, at his age I didn't know anything at all of Analysis, nor about electricity, though I had made several inventions (telegraphic) and was trying to see my way.' The four-year spell at Newcastle was indeed a period of intense searching but he did find his way and it turned out to be different from anyone else's.

It was not long before Oliver showed his mettle. Repairing a fault in the North Sea cable was a troublesome business because the cable ship's crew often started off with no more than a rough idea of where the fault was. To find out whether it was east or west of a given point they had to sail there, grapple the cable, haul it to the surface, buoy it, cut it and test it. After resplicing the cable they sailed to another point and went through the routine again. The process was repeated until they had narrowed the search down to a section of cable short enough to be economically replaced. Clearly, a lot of time could be saved if the fault could be approximately located before the ship left shore, and this is just what Oliver did. In the cable hut at Newbiggin-by-the-Sea he connected a battery to the cable and took two current readings, one with the far end of the cable short-circuited and one with it open. The rest was simply a matter of using Ohm's law – voltage equals current times resistance – and carrying out some algebra. He didn't know the resistance of the leaky fault but could use his other information to eliminate it from the calculation, leaving a quadratic equation that was easily solved to give an estimate of the position of the fault. Simple, but ingenious. Most 'practical' men of the time cared little for mathematics and the cable ship's crew probably took Oliver's estimate that the fault was 114 nautical miles from the English end with more than a pinch of salt. But they gave it a go and found he was right. This was cause for celebration and, for once, he let his hair down. Oliver's notebook entry reads: 'All over. Dined roast beef, apple tart and rabbit pie, with Claret, and enjoyed ourselves.' There was a touch of beginner's luck here – the method did not always work, because the fault's electrical resistance sometimes varied during the shore-based test – but he was off to a good start.

The notebook became an important accessory to his life. It served as a log for his experimental findings, as a sketchpad for his mathematical investigations and, on occasion, as a repository for his views on events and life in general. He was a compulsive observer of nature and one topic that intrigued him during his time in Newcastle was the weather – its causes and consequences. He made long notes on lightning and the aurora borealis and one entry reports that three recent earthquakes had darkened the sky with dust, so that it seemed the 'Day of Doom' had come.

Being back in England gave him the chance to resume his highly personal quest for knowledge. He was beyond mere 'book-tasting' now; he knew what he was looking for and the Public Library was his cornucopia. The scattered literary references in his papers and letters show that he read widely, Carlyle and Dickens being particular favourites, but at this stage what he sought most were books through which he could teach himself mathematics. He found two by people who had spent much of their lives teaching others. As we have seen, Oliver was not naturally receptive to orthodox teaching methods and he had to work hard to overcome his dislike of the way the subject was presented. He succeeded, but the experience reinforced his view that the standard way of teaching mathematics was wrong: it was really an experimental science and should be taught as such.

Even so, both books served their purpose well. One, by Isaac Todhunter, gave him a good introduction to calculus. Todhunter was a celebrated Cambridge tutor, a typical academic of his time and the very antithesis of the kind of scientist Oliver admired. One day in the street outside the Cavendish Laboratory in Cambridge, Todhunter had bumped into James Clerk Maxwell, who asked him in to see a rare example of conical refraction. Horrified, Todhunter replied: 'No. I have been teaching it all my life and I don't want my ideas upset by seeing it now.'

The other book, on differential equations, was by George Boole, creator of the system of symbolic logic which anticipated digital computers by the best part of a century but became their natural language. Heaviside called it that 'damnable book', but this did not mean he had a poor opinion of it. He always enjoyed a good-natured grumble; when he really took against people he used much stronger invective. In his view, Boole, like most mathematicians, gave too much prominence to logic: it had its place, of course, but that place was last – it was a tool for checking and tidying up once the creative work had been done.

Oliver had no interest in maths for maths' sake, but he had already begun to see that the only way to understand electricity and put it to best practical use was by harnessing the power of mathematics. He was years ahead of his time. The prevailing attitude of practitioners was neatly summed up by the weekly trade journal, *The Electrician*.

> Sciences generally become simplified as they advance, albeit in some branches they may intrude into the abstruse. In electricity there is seldom any need of mathematical or other abstractions; and although the use of formulae may in some instances be a convenience, they may for all practical purposes be dispensed with.

This was an editorial of ten years earlier, but things had changed little. Nor did they change for some time. Heaviside eventually shook the engineers out of their flat-earth mode but it turned out to be a long and bruising process. By a happy irony, it was *The Electrician* that later published many of his most abstrusely mathematical papers.

He had no way of knowing what a huge task lay ahead. To scale the vast, patchily charted mountain, he was eventually forced to make his own surveying instruments and climbing gear by adapting known mathematical methods and inventing new ones when needed. Meanwhile, he made forays into the foothills by experimenting with the kinds of electrical circuit that he had used in his work to measure voltage and resistance. Precision was important here and he found some clever ways to improve on the standard methods. From the first he felt it a duty to pass knowledge on to others, so he began to write up his findings and submit them to journals. His first paper, on a new way of comparing battery voltages, was published in the *English Mechanic* in July 1872. By the end of 1873 he had published six papers, four of them in the highly respected *Philosophical Magazine*, and his work began to be noticed. Among the people who saw it were three who came to play big parts in his life – two good, one bad.

Oliver's second paper drew the attention of William Thomson and James Clerk Maxwell, no less. He wrote in his notebook:

> 'On the Best Arrangement of Wheatstone's Bridge'. My first Philosophical Magazine Paper. A very short time after it appeared, saw Sir W. Thomson at Newcastle, who mentioned it, so I gave him a copy, which no doubt he didn't read. They say he never reads papers. Cuff told me Sir W. said he had tried to work it out, but found the algebra too heavy. ... So paper was a good beginning. Sent Maxwell a copy and he noted it in his 2nd Edn.

This has something of the air of a schoolboy cricketer who has scored a century but knows it is bad form to whoop and yell. He was thrilled. It was quite something for William Thomson, hero of the Atlantic cable and the most famous physicist in the country, to have noticed the work of a 22-year-old telegraph operator. Thomson's remark about finding the algebra too heavy was probably just a way of paying an encouraging compliment to the youngster: he may have dabbled at the problem on the train back to Glasgow after a tiring day in London but could certainly have cracked it if he had really tried. All the same, it was a tricky problem – to find what combination of five known resistances in the circuit will give the most precise measurement of the sixth, unknown, resistance – and it was a good test of Oliver's newly acquired knowledge of calculus, his algebraic skill, and his determination to keep plugging away until he got the answer out. Heaviside always had a high regard for Thomson, although he didn't always agree with him, and the feeling was returned: Thomson gave Heaviside solid support on several crucial occasions.

The contact with Maxwell was even more significant. From the moment he saw the great man's *Treatise on Electricity and Magnetism* in the Newcastle Library, Oliver was captivated. Many years later he recalled the experience.[1]

> I remember my first look at the great treatise of Maxwell's when I was a young man. Up to that time there was not a single comprehensive theory, just a few scraps; I was struggling to understand electricity in the midst of a great obscurity. When I saw on the table in the library the work that had just been published (1873) I browsed through it and was astonished! I read the preface and the last chapter, and several bits here and there; I saw that it was great, greater and greatest, with prodigious possibilities in its power. I was determined to master the book and set to work.

Mastering the book was to take him nine years. Even so, he was among the first to do it and he became the leading exponent and developer of Maxwell's electromagnetic theory. For the present, Oliver had good reason to be pleased with himself: he had earned a place in the greatest scientific book since Newton's *Principia*. Maxwell's *Treatise* has never been out of print; you can buy a copy today and see Oliver's result on the best set-up for the Wheatstone Bridge on page 482 of Volume I.

At about the same time, a dark side to Oliver's life started to form, although he knew nothing of it yet. It started with what one writer has described as 'perhaps the happiest day of his existence'.[2] Oliver had been thinking about the possibility of duplex working – using a telegraph line to send messages in both directions at the same time – and had devised some clever circuit arrangements for the purpose. Any telegraph company that managed to get double duty from its lines in this way stood to make huge profits, but most of the senior British engineers at the time thought that duplex was impractical – pie in the sky. Oliver thought otherwise and, as his notebook reports, he and brother Arthur made some experiments 'on duplex working with an artificial line and rough resistances at Beckett's shop'. We don't know who Beckett was or what he sold in his shop but let's hope he joined in the celebrations when Oliver's scheme for duplex worked like a dream. They even managed quadruplex operation, sending two messages simultaneously in each direction. Arthur then arranged for them to run a clandestine trial over real telegraph lines between Newcastle and Sunderland. More success: they sent messages 'simultaneously from both stations as fast as they could be transmitted by key'.

Elated, Oliver reported their findings in print. He did it under his name alone, fearing that Arthur's superiors at the Post Office would be none too pleased when they saw the article.[3] He was right. William Preece, Engineer in charge of the Southern Division, wrote to his boss, R.S. Culley:[4]

> Oliver Heaviside has written a most pretentious and impudent paper in the Philosophical Magazine for June. He claims to have done everything, even Wheatstone Automatic duplex. He must be met somehow.

Culley replied:

> O. Heaviside shows what is to be done by cheek. This we see every day – look at Thomson among the great, brings forward the tangent and sine scales on galvanometers as new. He does not read his Handbook it is evident. He claims or is supposed to have brought out lots of other things. We will try to pot Oliver somehow.

The senior engineers were beginning to marshal their forces against the brash upstart. Had Oliver made the conventional gestures of humility and respect for authority he might have become their blue-eyed boy. Instead, he ridiculed the head-in-the-sand attitude of the Post Office and said, more or less, 'Duplex is easy; I've done it.' He did not even acknowledge Preece's work on the topic – a patent taken out in 1855 and a paper in 1872 – and referred disparagingly to Culley's *magnum opus*, the *Handbook of Practical Telegraphy*. Had they charged him personally with these offences he would probably have replied that he was doing Preece a favour by not mentioning his scheme as it was badly conceived and would never work, and that Culley's *Handbook* was not really practical as it was hopelessly behind the times.

Diplomatic niceties aside, Heaviside was right on all counts. Preece's system didn't work; the *Handbook* was wrongly dismissive of duplex (this was corrected in the next edition); his own ideas on circuit arrangements for duplex were taken up in India and worked well; and duplex, even multiplex, operation later became commonplace. But he was already a thorn in the side of the engineering establishment and battle lines were being drawn for what became a long and bitter campaign.

Meanwhile, at the Great Northern Telegraph Company, business boomed and every available minute was taken up with telegraph traffic. Oliver still managed to incorporate some of his own investigations into the daily round, but the pioneering days of improvising and experimenting were passing, as things settled more and more into a routine. The main office in Newcastle became even busier when a second North Sea cable link was opened in August 1873, this one to Sweden. One group of operators punched outgoing messages on to paper tape and a second pasted tapes together to make continuous batches which they then transmitted. A third group decoded incoming messages and passed telegrams to waiting delivery boys. It was, on the whole, fairly mechanical and boring work once the skill had been acquired. But problems occurred from time to time and Oliver was in his element solving and fixing them.

One problem in particular puzzled everyone. It was to do with rain. Telegraphers had found that signals transmitted over long landlines became weaker in wet weather because some of the current leaked to ground, but the rate of signalling was not greatly affected. Over the Anglo-Danish link, however, the opposite happened – the strength of the signal hardly diminished but the pulses became even more smeared-out than usual, forcing the operators to send the messages more slowly in order to make them readable. Rain could make no difference to the undersea cable, and the landline on the English side was only 20 miles, so the natural suspect was the 120-mile line on the Danish side. It came in for a great deal of blame but Oliver knew it to be a sound line and sought the reason elsewhere. With tenacity and flair worthy of Hercule Poirot he tracked down the culprit. It turned out to be his uncle's invention, the Wheatstone automatic transmitter.

Wheatstone's transmitter was originally designed to send a positive pulse for each dot and a negative one for each dash; the received pulses would then be recorded on a paper tape by two rows of marks, one to represent dots and the other dashes. But a customer with a big potential order wanted actual dots and dashes, so Wheatstone had modified the design to suit. The modified machine used a more complicated system of positive and negative pulses to indicate dots, dashes and spaces. It worked well on landlines but ran into trouble on the Anglo-Danish link, where, in wet weather, it unbalanced the receiving device. This happened because the rain effect interacted with two other factors – the asymmetric lengths of the landlines and the intrinsically high capacitance of the undersea cable – in such a way that either the positive or negative pulses would not show up at the receiver unless the signalling rate was slowed down. All this was very complicated and Oliver was unable to persuade the company, or his uncle, that the machinery should be remodified. But it was the experience from this type of deep investigation that brought an unmatched tone of authority to his writings.

Not all Oliver's working time was spent in the office. Some days he made the 40-mile round trip to the cable terminal at Newbiggin to carry out tests on the Anglo-Danish cable. The job of maintaining the cable was contracted out to the firm of W. T. Henley but Oliver had a reputation as a fault finder and it seems that on at least one occasion they took him along when they went to work on the cable at sea. An entry in Oliver's notebook reads:

> The next few days were spent in waiting for Caroline. It was always leaving tomorrow. At last left and reached Shields. A week was spent in grappling, cutting and splicing cable.

This sounds like one of his good-natured grumbles – he clearly enjoyed the break from office routine and the chance to see the rugged side of the telegraph business. The crew of C.S. *Caroline* were a hardened bunch. They had not actually laid the Anglo-Danish cable but they had laid one across the Gulf of Bothnia from Sweden to Finland and had laid the eastern shore-end of the 1865 Atlantic cable at Valentia: the Newbiggin–Sondervig maintenance run was probably as near as they got to a home posting. Oliver was not the only electrician on the scene: Henley's had their own man, S.E. Phillips. There was obvious potential for friction but to judge from Oliver's cordial references to the older man in his writings, the two seem to have got on famously. Phillips was not one of the engineers who thought mathematics was only for ivory tower academics. Like William Thomson, he had tried to work out the optimum arrangement for the Wheatstone bridge but found the algebra too stiff, so Oliver came in for hearty congratulations when he cracked the problem.

By now the telegraph was a big industry and its engineers were important people. They felt the time had come to get the profession formally recognised, and formed the Society of Telegraph Engineers, with Sir William Siemens as president. Siemens was one of four brothers from Prussia who were all outstanding engineers. The eldest, Werner, had, as we have seen, invented the method of moulding gutta percha around a wire that made undersea cables a practical proposition. His brother Wilhelm had come to England at the age of twenty and liked it so much that he changed his name to William and stayed, becoming a pillar of the British engineering profession. Sir William Thomson agreed to serve on the council and the Society was off to a flying start – having the two famous men aboard helped to establish its status alongside other professional bodies and learned societies.

Membership also brought status to individuals and a sense of community, so it was natural for men like S.E. Phillips and Oliver's brother Arthur to sign up. Arthur thought that Oliver should join, too, so he made enquiries on his behalf. There was a brusque response – they didn't want 'clerks', the term then used for telegraph operators. Oliver was incensed; he knew his worth and suspected that the Post Office 'snobs' – Preece, Culley and company – were behind the rejection. On his next visit to London, emboldened by fury, he buttonholed Thomson and asked the great man to propose him. Thomson agreed at once but had engagements out of town so he asked Siemens to do the honours. No one could turn down a candidate proposed by the Society's president, so Oliver was in. He had beaten the 'snobs' at their own game.

Here, it seemed, was a young man set for a shining career in telegraphy, well regarded by his employers and already a published author of technical articles. And he was now in the club, a member of the professional association. But he was not at ease. There were echoes of his childhood: try as he might, he could not be like other people. As his views on life developed they diverged more and more from conventional tracks and there was nobody he could share them with. He found his soul companions in books, as a notebook entry reports:

> How true is Carlyle's observation that when a man, entertaining, as he believes, solitary heterodox opinions meets another holding the same opinions, how strongly is his belief strengthened! He is no longer alone! I have been reading today some of Tyndall's Essays on Materialism, Miracles, etc., and what was to be only a strong opinion is now a Conviction. But not unreasoning conviction.

John Tyndall was a good friend of Wheatstone's, a true meritocrat who had sprung from a modest background in Ireland to follow Faraday as Director of the Royal Institution. He was a fine scientific writer and his book *Heat as a Mode of Motion* was one of Oliver's favourites. Tyndall was one of the few top Victorian scientists – his friend T.H. Huxley was another – who made it known that they did not believe in God, or at least not in the kind of God represented in orthodox Church teaching. In some people's view this put him beyond the pale, but his words clearly struck a chord with Oliver.

By embracing a kind of scientific pantheism, Oliver set himself apart from his family, who were staunch Unitarians. Like Tyndall, he had come to believe that the road to true enlightenment lay through scientific understanding of nature, both living and inanimate, and he began to see himself as an evangelist in this cause. He was, as he later put it, 'a mighty enthusiast, filled with a strong sense of my Duty to impart my knowledge to others and help them'. His attitude to his job was ambivalent – he was keen and helpful on technical matters but increasingly resentful of the tedious hours of routine work. When working on the cable or its equipment he would muck in, doing whatever was needed, but in the office he refused to take a hand in the task of pasting paper tapes together, saying haughtily 'I am not a bookbinder'. Never one to undervalue himself, he thought that his contribution to the company was worth more than they were paying him and asked for a further pay rise. They were not used to such audacity from employees. As they saw it, Oliver Heaviside was already earning top money for a telegraph operator and, although clever and useful, was too self-willed to be a true company man. He was good, but not that good. The request was refused.

This result cannot have been a surprise but it was still a blow. The bid for more money could not have been just a try-on: Heaviside never went in for that kind of manoeuvring. He genuinely believed he was being paid less than his due, and may have been right: perhaps the bosses were not fully aware of his technical contributions. Whatever the rights and wrongs, the rejection drew into focus the matter of what he was going to do with his life. The right course was clear – his vocation was to find out all he could about electricity and spread the word. The six-year apprenticeship in telegraphy had given him as good an insight as anybody's into the ways of electricity, but a host of unanswered questions remained. The technology of the telegraph seemed to be in a rut. He felt sure that the best chance of progress would come from mathematical study of the fundamental workings of electricity but he had so far barely scratched the surface. It would be a full-time job.

What about money? He may have thought his salary unsatisfactory, but it was princely compared with the pittance he could expect from writing articles. He would scarcely be able to support himself, let alone help his ageing parents. Yet the prospect of self-imposed penury and financial dependence on his brothers seems not to have troubled him. There was none of the humiliation that most people would feel in this situation – he was doing what he saw as his duty and could hold his head high. He had no objection to money justly earned. If society chose to reward him with generous fees or royalties that would be fine; if not, his brothers and parents would be doing *their* duty by looking after him. As for starting a family of his own, the thought seems not to have entered his head. Although he enjoyed the company of women, he never, as far as we know, harboured romantic feelings. In all probability he knew himself to be far too egocentric to be a satisfactory lover, husband or father.

Before he finally cut personal ties with the business world, there was an offer from a surprising source. William Preece, who made it his business to know everything that went on, got to hear that Oliver was discontented at the Great Northern and offered him the possibility of research work for the Post Office, either in the Chief Engineer's research section or on his brother Arthur's own staff. Either way, he would need to report on his work to the Engineer-in-Chief, R.S. Culley, who would have control over what was published. Preece's motives were, no doubt, partly benevolent – he liked to bring on young talent and was probably prepared to forgive Oliver his impudent paper on duplex working in return for a show of deference. But there was self-interest, too, in attempting to tie down the loose cannon – in trying to 'pot Oliver', as Culley had put it. Oliver wanted none of this: the prospect of having his work directed and censored by people who didn't understand it was abhorrent to him. Arthur was in an awkward position as the man in the middle, wanting to protect both his brother and his own Post Office career. He resolved the difficulty by telling Preece, truthfully and with Oliver's agreement, that his brother was by now too absorbed in his own studies to be an effective member of a team.

In May 1874 Oliver resigned his post at the Great Northern, stating that he had found another situation. The new 'situation' was as an unpaid independent researcher into electricity, lodging with his parents in London.

No sooner had Oliver returned to London than he became very ill. What had begun in Newcastle as a chill became something far more unpleasant, affecting the digestive tract, the head and the nervous system. The condition was never properly diagnosed, but he called it 'the hot and cold disease' and suffered from it periodically for the rest of his life, together with ever-present dyspepsia. His mother, glad at least that he was at home, nursed him through it. She had suffered for much of her life from epilepsy, and from the strange effects his illness had on the nerves Oliver feared he might develop it, too. In fact, he never did, but it was a serious worry and from this time on he began to apply his own ideas on how to stave off attacks of the mysterious disease and minimise the effects when it struck. This was not hypochondria, at least not of the usual kind – he had no time for doctors and medicines – but it did lead to habits which seemed odd. He stuck to food which he knew suited him, and became obsessed with keeping warm – a visitor described his room in the house as 'hotter than hell'.

He fell into a way of living that had little connection with what most people regard as ordinary life. The day-to-day worries and pleasures that filled their lives had no place in his. It was a near-solitary existence, yet he seems to have had no feeling of loneliness. Later on he made some good friends but for now he had no one outside his family. Many years later he wrote of his life at this time.[1]

> There was a time indeed in my life when I was something like old Teufelsdröckh in his garret, and was in some measure satisfied with a mere subsistence. But that was when I was making discoveries. It matters not what others think of their importance. They were meat and drink and company to me.

Diogenes Teufelsdröckh was Thomas Carlyle's alter ego in his satirical book *Sartor Resartus* – an eccentric professor whose 'philosophy of clothes' provided metaphors for the author's own thoughts on life. Oliver's room was indeed like a philosopher's garret, but an unusual one, crammed with condensers, coils, galvanometers and other pieces of electrical apparatus. His meals were often left outside the door so that he could take them when he chose. During the day he read or made experiments and in the evening he would shut the window, light his pipe and immerse himself in theoretical work until the early hours of the morning. Exercise was not neglected. He walked for miles around London, taking in all that was going on – he was a keen observer of life even though he preferred to keep most of it at a distance.

For more serious exercise he went to a gymnasium at 'The Pimple', as Primrose Hill was known.[2]

Music was another favourite diversion. In a letter to a friend, Oliver wrote:[3]

> In old days I went to concerts, very long and highly classical; I always got wearied. I could not take it in – except the divine Schubert. Now there are a lot of very fine overtures of the Freischütz type. People hear them again and again, and get to know them. May their performance be never discontinued. ... I am very deaf. ... I have no knowledge (of music) nor am I a pianist, though I once taught myself B's Opus 90. I liked it better than anything else. Truly the conflict between the intellect and the heart.

There is a poignant irony here: Oliver may not have been aware that the composer of his favourite piece of music was himself almost totally deaf at the time he wrote it. Beethoven's Opus 90 is his Piano Sonata in E minor, a formidable work that takes about a quarter of an hour to perform. For anyone with no piano training to even think of attempting it is astonishing, and to have learned it well enough to produce a recognisable version must have taken enormous perseverance, especially as he found standard musical notation difficult. He confessed that the result was 'horrible to the musically trained ear' but felt he had captured the 'inner spirit' of the music and was pleased to have the achievement recognised by the family. Even his father complimented him – 'once only, but that was good for him'. We assume Oliver learnt the Sonata directly from a printed copy, because if he had transcribed it into his own musical notation he would certainly have told us. Unfortunately no clear examples of Heaviside's notation have survived, so we have no idea how he thought he could improve on the system of staves, bars and notes that everyone else used without complaint.[4]

In 1875, his parents moved again, to a slightly better part of Camden Town. Their new address was 3 St Augustine's Road and it was here, over the next 14 years, that Oliver produced a brilliant succession of startlingly original papers that eventually earned him the friendship of some of the world's best physicists. His work would come to have a profound influence on electrical communications and on the theory and practice of electrical engineering all over the world, but only after a period of slow diffusion. He did not help his cause. Not only were his papers hard to follow, sometimes unnecessarily so, but also he spurned a great chance to enlist support from the Society of Telegraph Engineers. In 1875 they appointed him to their Council, a remarkable honour for a jobless twenty-five year-old, but he didn't attend a single meeting. Worse than that, he failed to pay his subscriptions. Even then the Society renewed his membership for several years, but in 1879 their tolerance ran out and they expelled him. As with many aspects of Heaviside's life, we can only guess the reasons for his odd behaviour. Certainly his deafness would have made it hard to cope with large gatherings and he probably recoiled at the prospect of being patronised by such people as William Preece.

As to what daily life was like in his parents' house we have only a few scraps of evidence, and even in these one has to read between the lines. Nevertheless, the picture emerges of a household far more at ease with itself than it had been ten years earlier. The days of constant worry about money had passed, thanks largely to contributions sent by Arthur and Charles. Even Herbert helped out, though not entirely with good grace. To start with, Oliver had no income to contribute, but he may have chipped in with some savings from his time with the Great Northern. There were probably few outward displays of affection, but for all of Oliver's grumbles about his early life it is clear that he loved, admired and respected his parents. A couple of instances give us some idea of his relationship with his father.

In a notebook entry, he interrupts an account of an experiment to report how his father was wise to some leg-pulling.

> Father smells acid in the room. Two or three evenings. I said, at hazard, it was the electricity. Query, ozone generated by sparking, or nothing to do with it. Father says it is just like the battery he made when a boy, and that it is my battery. I didn't say it wasn't. What is the best arrangement to get the greatest variation of resistance in the circuit? I find that the internal and external resistances must be equal.

Another note shows how highly Oliver thought of his father's work, and of the way he had fought ill health to accomplish it. It was written in the margin of a proof copy of an engraving made by Thomas for an artist who had made a coup by getting Queen Victoria's permission to draw the elaborate interior of the newly completed Royal Mausoleum at Windsor where her beloved Albert was buried.

> This was done in less than a fortnight (12 days, I think) under great pressure from Mr Godwin when father … was very ill. £20. 1870. Franco-German War. (He asked £22.) Partly stress. Partly usual practice. Copy in bound volume also. Mr Godwin had sole permission from Her Majesty to give a picture. Owing to these circumstances, was very proud of it, but it would have been better to have had time to finish it finer.

Sometimes Oliver published reports of his experiments but they mostly served as a test bed for his theoretical work. Here, he had begun to feel the excitement of discovery and was eager to learn all he could from the experiences and discoveries of others. Maxwell's wonderful *Treatise* of a thousand densely packed pages was exactly what he needed on electricity and magnetism but much of it was stiff going. And he needed a broader base of knowledge to draw on: physicists like Maxwell and Thomson who made discoveries in electricity had often used analogies from other branches of physics. John Tyndall's *Heat as a Mode of Motion*, already mentioned, was useful and Oliver grumbled his way happily through William Thomson's and P.G. Tait's monumental *Treatise on Natural Philosophy*. This was the first proper textbook on physics and had a huge success at the bookshops. It became known simply as T and T′, the sobriquet Maxwell had given to its authors.[5] Oliver also discovered Jean Baptiste Joseph Fourier's superb *Analytic Theory of Heat*, which had already inspired many British scientists, including Maxwell and Thomson. Like them, he was smitten by Fourier's work, describing it as the only entertaining mathematical book he had read. By all this reading, he gained not only knowledge but an understanding of the process of discovery, complete with its wrong turns and blind alleys, and a feeling of fellowship with the men who had led the way. He felt he had joined the great club of natural philosophers, and later wrote of his disappointment that this traditional title had fallen into disuse.[6]

> For my part I always admired the old-fashioned term 'natural philosopher'. It was so dignified, and raised up visions of the portraits of Count Rumford, Young, Herschel, Sir

H. Davy, etc., usually highly respectable elderly gentlemen, with very large bald heads and much wrapped up about the throats, sitting in their studies pondering calmly over the secrets of nature revealed to them by their experiments. There are no natural philosophers now-a-days.

There was nothing of the ivory tower about all this study. Everything he read was assessed for usefulness and set alongside what he had seen for himself in the course of his practical work. And through Arthur he kept in touch with what was going on in the British telegraph industry.

The field of study was vast but at its heart lay the great mystery that captivated him: exactly how did electrical signals travel along wires? He went on to make the subject his own – he is to the transmission line what Thomas Hardy is to Wessex – and over the years built up the body of knowledge which is now taken for granted by electrical engineers. But in the mid-1870s nobody really knew how transmission lines worked. The only mathematical theory around was the one that William Thomson had dashed off in a few days 20 years earlier for the Atlantic cable. He had simply wanted to make an adequate mathematical model of how signals sent at a relatively slow rate would travel along a long undersea cable. For this purpose he needed to take into account only two properties of the cable: its resistance, which determined how much steady current would flow for a given voltage, and its capacitance, the capacity of the cable for storing electric energy. Thomson knew that transmission lines had other properties but decided that in the case of the Atlantic cable their effects would be small enough to be safely neglected. The mathematics was difficult and he didn't want needless extra complications. He knew that his results would be valid only for long undersea cables worked at low speed, but that didn't bother him. He was never really interested in overland telegraphy, at least not to the extent of getting personally involved.

But such was Thomson's standing that the main results from his theory became enshrined in the lore of telegraphers even though few, if any, of them could follow the mathematics. Practical men came to talk of the '*KR* law' and the 'law of squares' in the same way that they might mention the law of gravity. Unfortunately, they applied these laws outside their proper sphere and often didn't understand them anyway, with the result that some strange notions got about. Thomson's *KR* law for submarine cables said that the received signal was delayed by a time proportional to the product of the cable's capacitance and its resistance (they used K for capacitance in those days). It became misappropriated by William Preece, who used to tell everyone that the square root of $1/KR$, when multiplied by a constant of his own devising, gave the limiting distance for sending intelligible signals on any kind of line, K and R being respectively its capacitance and resistance per unit length. The law of squares followed directly from the *KR* law. It said that the maximum rate at which you could send signals – the number of letters or words per minute – varied inversely with the square of the length of the cable, but this had become widely misinterpreted as implying that the electricity itself moved more slowly in long lines than short ones. As we shall see in later chapters, Oliver made great sport of all this, swooping on the howlers like a gannet diving for fish. But for the moment he was absorbed with the task of picking up where Thomson had left off. To appreciate how difficult this task was we need to consider why it was that very few people really understood Thomson's theory.

All telegraphers knew about resistance. It was embodied in Ohm's law: if you apply a voltage V to a circuit with resistance R you get a current equal to V divided by R. But in telegraph transmission Ohm's law applied only in the steady state, once everything had settled down – the important question was what happened *before* everything settled down. When you pressed or released the sending key, the transient relationships between voltage, current, time and distance were far more complicated than Ohm's law.

Something similar applied to capacitance. Many people thought they understood it because one type of capacitor, the Leyden jar, had for a century been a popular source of parlour entertainment: it stored static electricity and could be made to release all its charge on command, thereby setting off a show of sparks or giving people shocks. But working a telegraph line was not like charging or discharging a Leyden jar. What mattered was not just how much electricity the line stored when it was fully charged but, more importantly, how the line behaved while it was charging or discharging. Thomson had derived the equations for voltage and current during these transient periods for the simple case of a line with only resistance and capacitance, but the mathematics was beyond the reach of all but a few élite physicists and mathematicians, so it is not surprising that his results were poorly understood.

Despite having no one to guide him in his study, Heaviside mastered the mathematics and set about improving Thomson's theory. Besides resistance and capacitance, which were at least partly understood, electrical circuits had another property that was altogether more mysterious – electromagnetic induction – and Oliver's first big breakthrough was to bring it in to the mathematical analysis and show that it had a crucial effect on the way transmission lines worked.

Electromagnetic induction was Michael Faraday's greatest discovery. Much of our present-day technology depends on it but, even now, it is not easy to understand. As Oersted had shown in 1820, an electric current exerts a magnetic force. Faraday thought this happened because the current produced a magnetic field – a system of 'lines of force' filling all space. If electric currents produced magnetism, surely magnetism could be made to produce electricity. Scientists tried everything they could think of but all attempts failed until, in 1831, Faraday found that an electromotive force, or voltage, could be induced in a circuit by *changing* the number of magnetic lines of force that *passed through it* – the faster the change, the greater the voltage.

For example, moving a magnet towards a nearby loop of wire (or vice versa) increased the number of lines of force passing through the loop and thus induced an electromotive force which made a current flow in the wire. But the current's own magnetic field opposed the movement of the magnet, so to generate the current you had to *push* the magnet, using mechanical force. The energy you supplied by pushing the magnet had to go somewhere: it went into the magnetic field. By the same token, when a battery was first connected to a circuit it had to supply the energy needed to build up the current and this energy went into the magnetic field surrounding the wire.

So a transmission line held two types of energy: electric energy through its capacitance and magnetic energy through its inductance*. Whereas the energy of the electric field in a charged cable resembled that of a stretched spring, the energy of the magnetic field set up by the current was rather like that of a flywheel. The electric field had a kind of elasticity and the magnetic field a kind of inertia. When a telegraph operator closed his sending key, the first flow of energy from the battery went into stretching the electric spring and getting the magnetic flywheel up to speed; only then would a steady current flow to the receiver. And when the key was opened again, the current did not stop at once; it continued until the spring had relaxed and the flywheel had stopped.

The problem facing Heaviside was to find the equation that gave the voltage at any given distance along the line at any instant; the solution would then also yield the corresponding equation for the current. The way to tackle this kind of problem was (and still is) to do it in two stages. First he had to find the right differential equation – the one that correctly represented the rates of change of voltage with distance and time. The task then was to solve this equation – to convert it to one that gave the actual voltage, rather than its rate of change, for a given set of circuit conditions.

Thomson had tackled a similar problem 20 years earlier but, as we have seen, he had simplified things by ignoring inductance. His differential equation, amazingly, turned out to have the same form as one that Jean Baptiste Joseph Fourier had put forward half a century earlier to represent the flow of heat in a metal bar. To solve this and similar equations, Fourier came up with the brilliant idea of replacing an awkward mathematical function by an equivalent infinite series of more amenable ones. With Fourier's example to help him, Thomson worked out the cable's response to a Morse code dot, a short pulse keyed in at the sending end. He found that the pulse became smeared out, which is indeed what telegraphers had observed, and his results were summarised in the famous, and much misused, *KR* law and law of squares.

Heaviside's task was harder, because he had to bring inductance into the reckoning, but he handled it with aplomb and wrote up the results in a paper with the arresting title 'On the Extra Current'.[7] In his hands the transmission line came alive: it was no longer a mathematical cousin of a heated metal rod. Using Fourier's method, Oliver investigated how the line would respond if the operator pressed the sending key and kept it down. What he found was that under some conditions the voltage and current would *oscillate* before settling down to steady values. Similarly, when the key was released, they would take time to revert to zero and would swing back and forth while doing so. So the current would perform a little dance of its own after the battery had been disconnected; this was the eponymous 'extra current'.

While on the topic, he couldn't resist seeing what would happen if, when the sending key was released, the line was simultaneously opened at the receiving end. The line would then be electrically isolated, connected to neither battery nor earth.

---

* The inductance of an electrical circuit is a constant which determines the amount of magnetic energy stored by the circuit when a given current flows in it: the higher the inductance the higher the energy.

But, even with nowhere to go, the current would, if conditions were right, continue to flow, out and back, until all its energy had been dissipated as heat.

The oscillations happened because energy was being shuffled between the line's electric and magnetic fields – from the electric spring to the magnetic flywheel and back again. What stopped this going on for ever was the line's resistance, which acted as a kind of friction, damping out the oscillations by converting the energy into heat. The higher the resistance the quicker the oscillations would die down. Too much resistance, and the oscillations would be stifled at birth.

The 'extra current' was not entirely unknown to telegraphers. They had some-times seen unexplained oscillations on landlines or when testing submarine cables coiled on their great reels before being laid at the bottom of the sea. Heaviside had now shown that these observations were not anomalies caused by freak conditions or faults in the apparatus; they were exactly what was to be expected when the cir-cuit conditions brought electromagnetic induction into the picture. According to his equations, oscillations were most likely to happen on lines where the capacitance was relatively low and the inductance high, for example landlines using receivers with strong electromagnets. And although submarine cables had high capacitance, their inductance increased enormously when they were coiled, so the conditions for oscillation applied there, too.

The first telegraphers saw their wires as nothing more than inert conduits, like water pipes, along which the battery pumped a kind of fluid called electricity. Things had come a long way since then – according to Heaviside the transmission line was a complex piece of equipment with properties that resembled elasticity, inertia and friction. He was way ahead of his contemporaries in understanding how electrical signals passed along wires, but the picture still lacked three main features.

In fact, two of these were already within his grasp, although he hadn't yet recog-nised their full significance. One was the way that capacitance and inductance acted together to transmit signals as *waves* along the line. The other was leakage: even when the line had no faults, some of the current did not get to the far end but leaked to earth through the insulating material. This weakened the signal but, strangely, it also turned out to bring tremendous benefit by making possible one of Heaviside's greatest gifts to the world – the formula for a *distortionless* transmission line. As we shall see in a later chapter, his brainchild made national and international telephone networks possible but, at the same time, caused bitter disputes on both sides of the Atlantic.

The leakage property was nothing new to Heaviside. He knew more about faults (local leaks) in cables than anybody and wrote a long and detailed paper about them, showing mathematically how they sometimes sharpened the signals and allowed more words to be sent per minute. He even suggested that signalling would be improved by building a deliberate 'fault' into a cable, halfway along its length and having 1/32 of the resistance of the cable. He was right, but the idea seemed mad and no one took it up. In another paper he took account of the small but ever-present leakage all along the line, together with the line's resistance and capacitance. But the full power of his analysis of the transmission line, together with the formula for eliminating distortion, only became apparent when all four properties – resistance, capacitance, inductance and leakage – were considered together, and it took him several more years to discover this.

In fact, he was already close to finding the wonderful formula for ridding lines of distortion – so close that it is tempting to think along the lines: 'All he had to do was so and so; how did he miss it?' But hindsight is sterile, and to think that way would be like looking through the wrong end of a telescope. It is not so hard to find a needle in a haystack if you know exactly where to look. Heaviside had the whole stack to search and didn't even know he was looking for a needle.

We shall see later how he found it, and also how he wove the third of the three new features into the tapestry. This was Maxwell's theory of electromagnetism, by which Heaviside showed that real transmission lines behaved in a way that not even he could otherwise have imagined. For example, the energy of the signals flowed not through the metal wires but through the space or insulating material surrounding them. The only energy flow inside the wires themselves was inwards, from the surrounding space, and this was just the portion of energy that was wasted as heat!

In the course of investigating transmission lines in all possible conditions and configurations, he cleared up an unsolved puzzle from his days with the Great Northern: why could eastbound signals be sent at a higher rate than westbound ones? This strange phenomenon had flummoxed everyone. It seemed that the difference in the length of the landlines – 20 miles on the English side and 120 miles on the Danish – must have something to do with it, but nobody could work out why this should favour eastward, rather than westward, transmission. The surprising answer was that the internal resistances of the terminal equipment – the battery at one end and the receiver at the other – were just as important as the much larger resistances of the two landlines. The reason for the eastbound signalling rate being quicker turned out to be mind-bogglingly intricate. Had *either* the landline resistances been equal *or* the battery and receiver resistances been equal, then maximum signalling rates would have been the same in both directions. Eastward signalling was quicker because the receiver had more resistance than the battery, so that when sending eastward the sending end had *both* lower landline resistance *and* lower terminal resistance than the receiving end. If the resistance of the receiver had been less than that of the battery then westward signalling would have been the quicker![8]

Actually, the solution had needed no more than Thomson's 30-year-old theory of the cable and some elementary circuit theory. Why then had it been so elusive, and why did Heaviside succeed where others had not? One reason is that circuit theory was still thought to be strange country into which you ventured at your peril. The days when it would become second nature to electrical engineers were far off. There is a great paradox here. Heaviside was the main architect of the transformation – in a sense it was his life's work – but time and again he left his readers to sink or swim, stubbornly refusing to take the trouble to help them understand his papers. In presenting his solution for the Anglo-Danish link, for example, he didn't set out his reasoning step by step but simply stated the long formula for the cable charging time with no explanation whatever. Instead of being enlightened, most readers were bewildered. This obtuseness is all the more puzzling because he believed passionately in the duty of a scientist to share his findings with the world – he considered Henry

Cavendish's failure to do so a criminal act.[9] When his own thinking ran along lines comprehensible to those less gifted, he was capable of the most brilliant teaching – demonstrating complex-seeming mathematical theorems by physical interpretations that made them almost self-evident. But such examples are few. For the most part he remains an enigma – as his close friend G.F.C. Searle put it, 'a first-rate oddity'.

Many of Oliver's findings could have been of immediate practical value to the telegraph industry but they were hidden in a forest of dense and difficult mathematics and, for the most part, passed unnoticed. But the benefits from the work, when they came, were longer-term and wider. One was the genesis of electrical circuit theory as it came to be taught and practised everywhere. Heaviside never published a full tutorial on the topic, and the ideas came out piecemeal, but the central theme was the brilliantly simple one of generalising Ohm's law. Everyone knew Ohm's law – voltage equals current times resistance – but in general it applied only when voltages and currents were steady and so was of limited use in applications such as telegraphy which dealt in pulses. When voltage and current varied, it was no longer sufficient just to consider the resistance of the circuit: its capacitance and inductance had to be brought in too, but nobody knew how to do it. Heaviside showed that capacitive and inductive elements in a circuit could be combined algebraically just like resistances as long as they were expressed in the right way, as 'operators'. This was the start of his operational calculus, of which more later. For a single resistive element, Ohm's law said that the voltage across it, V, was equal to its resistance, R, multiplied by the current flowing through it, I. Simply:

$$V = R\,I$$

Oliver's inspired idea was to regard R not just as a number but as an *operator* that converts a current into a voltage. He then read the equation as: R operating on I gives V. This was the key that opened everything up. The corresponding equation for an inductive element, of inductance L, was, when written in the normal way:

$$V = L\,dI/dt$$

where dI/dt means the rate of change of the current, I.

In Heaviside's interpretation, the operator now acting on I to give V was $L\,d/dt$, or L multiplied by rate of change of... Of what? Of whatever was being operated on, in this case the current, I. Strange enough, but he went further by giving d/dt the single symbol p and treating it algebraically just like a number. Capacitive elements were handled in a similar way, also using the symbol p. This way, even a complicated circuit could be represented by an operational equation that was really a differential equation but looked like an ordinary algebraic one and could be rearranged, using the normal rules of algebra, into a form from which it could be solved.

A remarkable early product of this idea was the method used today for analysing power circuits, in which both voltage and current oscillate at a fixed frequency. Heaviside showed how to do it – though not yet with the p notation – in 1878. This was about 15 years before AC power systems came into use.[10]

The same paper contains his first reference to 'the Bell telephone', which he called 'This most sensational application of electricity'. Alexander Graham Bell was a young Scottish-born expert in voice physiology who lived in Boston, Massachusetts. Almost to the month that Oliver's paper 'On the extra current' was published, Bell took out a patent on 'The method of, and apparatus for, transmitting vocal and other sounds telegraphically'. The telephone was, indeed, a wonder. Even to regular users of the telegraph, the notion of *talking* to remote friends or colleagues over wires had seemed a fantasy, but Bell's instrument was so simple that it could have been produced 30 years earlier.

His idea was: (1) use the sound waves from the speaker's voice to vibrate a metal diaphragm; (2) use the vibrating diaphragm to vary a magnetic field; (3) use the varying magnetic field to generate a similarly varying current in the telephone wire; and (4) use a similar device in reverse at the far end to reproduce the original sound waves. Now that Bell had opened the door, Thomas Edison and others were quick to improve on his design and to put their own products on the market. Commercial operations started up and the telephone was on its way, slowly at first, to becoming as natural a part of everyday life as walking the dog.

Oliver's ground-breaking work on transmission lines applied even more strongly to the telephone than to the telegraph, because the demands on the line were much heavier. To transmit understandable speech the line had to be able to handle frequencies of up to several thousand cycles per second, whereas a few hundred cycles per second had been enough for telegraph pulses. The higher frequencies led to a problem with cross-talk. When lines were close to one another their magnetic fields become coupled, like flywheels geared together, so that changes of current in one circuit induced changes of current in the other and you heard other people's conversations as well as your own.[†] But the crippling problem was distortion. Not only did signals become weaker as they went down the line, they also lost their shape because the high frequency components were more weakened than the lower. And the longer the line, the worse the distortion became.

The big telegraph organisations were naturally keen to bring the telephone into their schemes and quickly established local networks. But the problem of distortion limited the range of decent quality transmission to around 100 miles at best. The new technology held out the prospect of huge profits but it had run into a wall. Distortion barred the way to riches beyond compare, and it seemed that a fortune lay in store for anyone who could show how to get rid of it. As we shall see, Oliver did indeed show the way, but someone else made the fortune.

[†] The effect of the coupling of the magnetic fields of two neighbouring circuits was called mutual induction, to distinguish it from the self-induction of each circuit on its own.

Chapter 5
Good old Maxwell!
London 1882–86

Oliver went on ploughing his lonely furrow, examining electrical circuits from all angles and writing up the results for publication. Sometimes he reported on experiments, for example giving admirably clear guidance on the use of carbon contacts in microphones, but most of the papers were theoretical and packed with mathematics of ever-increasing complexity. His well of inspiration was full and his main concern was to find enough outlets for the words and symbols that poured from his pen. By his thirtieth birthday he had published fifteen papers but they were spread rather haphazardly among five different journals. This way, he was able to keep the traffic flowing without serious log-jams but he could never be sure that successive papers with a common theme would appear in the same journal. Then events took a happy turn. In September 1882 he had a letter from C.H.W. Biggs, editor of *The Electrician*.

> I should be greatly obliged if you would let me have a paper or two for The Electrician. Pray choose your own subject. The sooner I can have them the greater will be my obligation to you.

Can a duck swim? Biggs soon had his response and wrote back:

> Many thanks for your recent communications, which are just what I want … So long as you remain good to write, so long shall I be pleased to receive and insert your MS … I may say that I hope this will be a period of infinitely long duration.

It was indeed the start of a long and fruitful association. *The Electrician* was a weekly trade journal for engineers and, on the face of it, an unlikely showcase for Heaviside's work. But it was very well produced and had carried contributions from top scientists, including Thomson and Maxwell. It published most of Heaviside's papers for the next 20 years and this gave him a steady income of about £40 per year, almost enough to pay the rent on his parents' house.[1] But this happy outcome had a less pleasant aspect. Although he was no expert in mathematics, Biggs had been impressed by the authoritative tone of Heaviside's work and felt sure that here was something new and important. But some members of *The Electrican*'s editorial board were by no means as enthusiastic and Biggs knew he was going out on a limb by taking on his new regular contributor.[2] He never met Oliver but it was not for want of trying; he begged him time and again to come to the magazine's annual dinner, where the great and good of the electrical establishment met. Biggs must have known that Oliver had already ruffled some eminent feathers and may have hoped to bring him back into the fold. Oliver declined the invitations, and any hope of reconciliation with his arch-enemy William Preece was dashed five years later when Preece suppressed one of his papers. Although not directly involved, Biggs became caught up in the row that followed and could have made his life a lot easier by ditching Heaviside. When he took the honourable course – sticking by his writer and, by the same token, to the ethics of his profession – it cost him his job.

More of this later, but the very first paragraph of Oliver's first article must have given Biggs some idea of what he was letting himself in for.

> The daily newspapers, as is well known, usually contain in the autumn time paragraphs and leaders upon marvellous subjects which at other times make way for more pressing matters. The sea serpent is one of these subjects.

He went on to make sport of a press report about a boy with a stick who could detect water deep underground, and only then got down to the business of the paper, which was a masterly short account of a strangely neglected topic, 'The Earth as a Return Conductor'.[3]

This was just the beginning. Heaviside hated any kind of pomposity and, feeling secure in his new arrangement with *The Electrician*, he continued to lace his papers with whatever light-hearted notions took his fancy. They must have raised a chuckle in the editorial office because Biggs printed many passages which more cautious editors would have cut out. Something Oliver really enjoyed was pricking dignity. No person or institution was above some playful debunking and his sallies sometimes got him (and his editor) into trouble. One of his favourite targets was the Church. Darwin's *Origin of Species* had provoked a counter-attack by some members of the clergy, who claimed to see evidence of divine creation everywhere, and this was too good an opportunity to be missed. He followed up an earnest account of the physical principles of Ohm's law with a mock-homily on creation.[4]

> I cannot help unburdening myself of a conviction that has long been forcing itself on my mind with ever-increasing persistency. Ohm's law ... long ago became of immense practical importance, owing to the commercial value of the applications of electricity. But in the last few years its use has been multiplied enormously, through electric lights and one thing and another. In fact, for once it is used by the telegraph engineer it is used twenty times or more by an electric light man. Conceive then if you can, the electric light young man having to consult at every step volumes of tables and intricate formulae to ascertain what current corresponds to a given E. M. F.* in a given wire under given circumstances if its resistance were not, as it were, practically constant; if it were to vary, for instance, with how long the current had been on, or with the E. M. F., or in innumerable other ways; for who shall say what might not have been? The electrician would be in a perpetual state of embarrassment, which would certainly tend to shorten his days though it lengthened his day's work. Now is this fitness of things as regards Ohm's law not a powerful argument in favour of design? Can it be doubted that the constancy of R was providentially arranged to meet the requirements of man? I think this can only be contested by those wicked writers of dreary articles in monthly magazines whose pernicious stuff appears to be deliberately intended to undermine all our inherited faiths, and who seem determined to leave no place in the world for its own maker.

---

* Electromotive force, or voltage.

'Conceive, then, if you can, the electric light young man' was a variation on a popular song from Gilbert and Sullivan's *Patience*. One of *The Electrician*'s readers, Professor George Minchin of the Royal Indian College, picked up the satirical theme and wrote in about an archbishop who had claimed in a sermon to see the hand of the Creator in all science, citing the electroscope as evidence. Biggs published his letter and it was enough to set Oliver off again. He wrote back:

> As to Professor Minchin's remarks on the archbishop, I must say that I do not think them quite respectful, I really do not. Archbishops are privileged. They have reached the highest summit of respectability in an ultra-respectable country, to say nothing of their powers of intercession to turn a drought into a damp. I should hardly dare to criticise an archbishop, not if he fired off a whole volley of electroscopes and Leyden jars in defence of the Faith. I had the great pleasure of seeing a live archbishop this summer, the new one, entering Westminster Hall, no doubt on his way to his parliamentary duties, perhaps to stop men from marrying their deceased wives' sisters, and a nicer looking man for an archbishop I could not conceive. He wore the most delightful pair of trousers, buttoned tight all the way up the calves, and my respect for the cloth, always great, was much enhanced thereby.

Perhaps Biggs should have nipped Heaviside's and Minchin's schoolboy jape in the bud. Among those upset by it was one of the last people in the world Oliver would have wished to offend, his erstwhile champion William Thomson. A rueful entry in the notebook records Oliver's dismay on finding this out several years later. It ends with the comment: 'Really, however, there was nothing to be disgusted about. It was simply a bit of fun.'

This episode brings into sharp relief two contrasting elements of our hero's character. His puckish sense of fun could be beguiling, but he infuriated many people by refusing to make allowance for the feelings (or failings) of others. It seems that he simply could not, or would not, bring himself to imagine how the world seemed from someone else's point of view. Both elements were to play a part in dramatic scenes to come.

Bit by bit, Heaviside had been working his way through Maxwell's thousand-page *Treatise*, in which the author had set out to help others by describing all that he had himself learned or discovered about electricity and magnetism, both theory and practice. One of the few topics it didn't deal with explicitly was the telegraph. It was not intended simply as a showcase for Maxwell's own theory of electromagnetism; the theory was there, but you had to hunt for it. At first, Oliver was as perplexed as anyone by Maxwell's theory, but once he had interpreted it in his own way he was completely won over and became a lifelong evangelist in Maxwell's cause. Towards the end of his life, he wrote a moving tribute to the great man.[5]

> A part of us lives after us, diffused through all humanity – more or less – and through all Nature. This is the immortality of the soul. There are large souls and small souls ... That of a Shakespeare or Newton is stupendously big. Such men live the best part of their lives after they are dead. Maxwell is one of these men. His soul will live and grow for long to come, and hundreds of years hence will shine as one of the bright stars of the past, whose light takes ages to reach us.

When Maxwell died in 1879, aged only 48, very few people who worked with electricity even knew of his theory, let alone understood it. But to Heaviside it was a revelation – a completely new way of thinking that brought everything together and made sense of it. It became his constant star: if an idea was consistent with Maxwell's theory it was worth pursuing; if not it was given short shrift. 'Good old Maxwell!' he would say. He felt compelled to spread the word and set out to do so in his series of articles in *The Electrician*.

Heaviside was one of the first to recognise the importance of Maxwell's theory but even he might be surprised at the place it now holds as a cardinal law of nature, one of the central pillars in our understanding of the universe. It opened the way to the two great triumphs of twentieth-century physics, relativity and quantum theory, and survived both of those violent revolutions completely intact. As a bonus, it brings us radio, television, mobile telephones and radar. At the time, however, it was merely one of several competing theories. What made it different was the idea of the *field*, which had started with Michael Faraday's notion of lines of force.

To explain electric and magnetic forces, mathematical physicists such as André-Marie Ampère and Wilhelm Weber had assumed that electric charges and magnetic poles acted on one another *at a distance*, and that nothing happens in the space between them. Michael Faraday, on the other hand, believed that charges and magnets infused all space with 'lines of force', which interacted with the lines of force emanating from other charges and magnets to produce the electric and magnetic forces that we feel. In his view the space surrounding charges and magnets was not passive but was itself imbued with a kind of tension that gave rise to the forces.

Many people thought Faraday's ideas fanciful and for some years the 'action at a distance' school held sway. The British Astronomer Royal, Sir George Biddell Airy, put the prevailing opinion rather bluntly:[6]

> I can hardly imagine anyone who practically and numerically knows the agreement [between calculations based on action at a distance and experimental results] to hesitate an instant between this simple and precise action on the one hand and anything so vague and varying as lines of force on the other.

But not everyone agreed. Two young Scottish physicists found ways to show that Faraday's ideas were, at root, highly mathematical, even though their author, having no mathematics himself, could express them only in words. William Thomson discovered the equations for electric lines of force while he was still an undergraduate at Cambridge. They turned out to have the same form as those found by Fourier for the diffusion of heat through a solid material and, as we have seen, Thomson put the analogy to good use when working out his theory for the Atlantic telegraph. This was a fine start, but it was Thomson's young friend James Clerk Maxwell who really got to grips with Faraday's ideas. In three stages, spread over nine years, he transformed Faraday's notion of lines of force into the mathematically impeccable electromagnetic field and built a theory that changed the way physicists look at the world.

The whole route was paved with inspired innovations, but one crucial step stands out: the prediction that fleeting electric currents exist not only in conductors but in all materials, and even in empty space. There was no experimental evidence for this extraordinary idea; nor was it prompted by logic. But it was the missing link that brought all the laws of electricity and magnetism together into a compact and beautiful theory. It also meant that every time a magnet jiggled or an electric current changed, a wave of energy would spread out into space, like a ripple on a pond. Maxwell calculated the speed of the waves and it turned out to be the very speed at which light had been found to travel. At a stroke, he had united electricity, magnetism and light. And visible light was only a small band in a vast range of possible waves, which all travelled at the same speed but vibrated at different frequencies.

From our viewpoint almost a century and a half on, the fact that Maxwell had calculated the speed of light solely from the properties of electricity and magnetism seems to be overwhelming circumstantial evidence that he was correct. Even though experimental confirmation did not come for another quarter of a century, one may wonder why his theory did not immediately sweep all others away. Probably the main reason was that it was so different from anything that had gone before that most of his contemporaries were bemused. They simply did not know what to make of it. What made things even harder for them was that Maxwell had presented his results in two quite different ways. In the first, in 1861 and 1862, he explained how he had worked everything out using a bizarre imaginary mechanical model, in which all space was filled with tiny spinning cells with even smaller particles acting like ball-bearings between them. By giving his little cells both inertia and springiness he not only explained electric and magnetic forces but predicted that any changes in either would send ripples of energy into space – electromagnetic waves.[7]

The idea of an 'aether' – a medium pervading all space – was far from new. Physicists believed an aether of some kind was necessary to transmit light waves, so one might have expected ready acceptance of Maxwell's extension of the idea to electricity and magnetism. But there were misgivings: the reaction of his friend Cecil Monro was typical:

> The coincidence between the observed velocity of light and your calculated velocity of a transverse vibration in your medium seems a brilliant result. But I must say I think a few such results are needed before you can get people to think that every time an electric current is produced a little file of particles is squeezed along between two rows of wheels.

What people could not understand was that Maxwell's model was simply a thought experiment. They thought that all physical phenomena resulted from mechanical action of some kind and that all would become clear to us if, and only if, we could discover the true mechanisms. To most of Maxwell's contemporaries, the model was an ingenious but flawed attempt to portray nature's machinery. Even the most enlightened of them thought that the next step would be to continue the search for the true mechanism by refining the model. But Maxwell astonished everyone again by putting the model on one side and building up the whole theory from scratch, using only the laws of dynamics. In what must have seemed like a conjuring trick, he didn't worry at all about the detailed mechanism but treated it like a 'black box'. Using the methods of the French mathematician Joseph-Louis Lagrange, he managed to work out the relationship between what went into the box and what came out, without assuming anything about the internal mechanism except that it obeyed the laws of dynamics. This was enough to give the whole theory, electromagnetic waves and all.

Perhaps Maxwell had foreseen what twentieth-century physicists came to realise: that nature's ultimate mechanisms may be beyond our reach. But this was all too much for the audience at the Royal Society in 1864 when he presented his paper, 'A Dynamical Theory of the Electromagnetic Field'. Even among those who could follow the mathematics, some thought that abandoning the mechanical model was a backward step. Among them was William Thomson, who, for all his brilliance, never came close to understanding Maxwell's theory.

Heaviside not only understood the theory but re-expressed it in a form that was much easier to grasp, reducing the number of equations from twenty down to four. These are, in fact, the four famous 'Maxwell''s equations'. Although they are implied in his work, Maxwell never presented them together in the form now used. They are, in part, Heaviside's creation, although he rarely gets the credit today. But, as we shall see, his work on Maxwell's theory did bring him a measure of fame and some good friends. One of them later described Heaviside's achievement.[8]

> Maxwell, like every other pioneer who does not live to explore the country he opened out, had not had time to investigate the most direct means of access to country nor the most systematic way of exploring it. This has been reserved for Oliver Heaviside to do. Maxwell's Treatise is encumbered with the debris of his brilliant lines of assault, of his entrenched camps, of his battles. Oliver Heaviside has cleared these away, has opened up a direct route, has made a broad road, and has explored a considerable trace of country ... The maze of symbols, electric and magnetic potential, vector potential, electric force, current, displacement, magnetic force and induction, have been practically reduced to two, electric and magnetic force. Other quantities, it may be convenient, for the sake of calculation, to introduce, but they tell us little of the mechanism of electromagnetism. The duality of electricity and magnetism was an old and familiar fact. The inverse square law applied to both, every problem in one had its counterpart in the other. Oliver Heaviside has extended this to the whole of electromagnetism.

How did he do it? His answer would probably have been 'by hard work', and indeed it was. But the key to it all was his creation of a new language for investigating and describing electric and magnetic forces and the way they interact with one another. It suited its purpose so well that it has become the standard language that physicists and engineers use for dealing not only with electromagnetic fields but with other topics such as fluid dynamics. Now called simply vector analysis, it seems so natural a part of the scene that almost nobody wonders how or when it came about.[9]

Vectors like electric and magnetic force were different from ordinary 'scalar' quantities like mass or electric charge because they had both magnitude and direction. And because they operated in three-dimensional space, the usual way of representing them mathematically was by their components in each of three directions at right angles to one other. So even the simplest relationship between vectors needed three equations to describe it, and the physical meaning became obscured in a forest of symbols: one could not see the wood for the trees. Heaviside cut away the tangled vegetation and found a way to represent and analyse relationships between vectors using a single symbol for each. The resulting vector equations were not only simpler; they got to the heart of the matter by describing the intrinsic physical relationships independently of any coordinate system.

His preferred method for deriving equations was to map out the vector fields in his mind, resorting to symbols only at the end, when he had worked everything out. This demanded remarkable powers of visualisation because the geometry of vectors is quite unlike the ordinary geometry of shapes, lines and angles. For example, at any point in the field a vector has two key properties that describe its variation in space: the vector called curl and the scalar called divergence. We can get a rough idea of these concepts, and even of the four great 'Maxwell's equations', by following Heaviside's lead and trying to form a mental picture of the way that fields behave.

Think of water flowing in a river. The vector here is the rate of flow – both speed and direction – and, in general, it varies from point to point in the river. Imagine a tiny paddle-wheel somehow fixed at a given point, but with its axis free to take up any angle. If the water flows faster on one side of the point than the other, the paddle-wheel will spin and its axis will take up the angle that makes it spin fastest. The curl of the water flow at our point is a vector directed along this axis with magnitude proportional to the paddle wheel's rate of spin. In more general terms, curl is a measure of the field's vorticity, or swirl, at a given point.[†] It is at the heart of the relationship between electricity and magnetism. Two of the four equations in which Heaviside summed up Maxwell's theory show exactly how electric and magnetic forces fit together: at a point in empty space the curl of each is proportional to the rate at which the other changes with time. The equations are:

$$\text{curl } \mathbf{E} = -\mu \partial \mathbf{H}/\partial t \qquad \text{curl } \mathbf{H} = \varepsilon \partial \mathbf{E}/\partial t$$

where $\mathbf{E}$ and $\mathbf{H}$ are the electric and magnetic force vectors, $\partial \mathbf{E} \partial t$ and $\partial \mathbf{H} \partial t$ are their rates of change with time, and μ and ε are the fundamental constants of magnetism and electricity.[10]

We just have to look at the two equations to get an idea of how they act together to make waves in space. A changing magnetic force wraps, or curls, an electric force $\mathbf{E}$ around itself. As that electric force changes, it wraps itself in a further layer of magnetic force $\mathbf{H}$, and so on. Thus the changes in the combined field spread out in a kind of continuous leapfrogging action; they are electromagnetic waves. Their speed works out to be $1/\sqrt{(\mu\varepsilon)}$, which, with values inserted and reasonable allowance for experimental error, turns out to be the very speed at which light had been measured. In other words the speed of light depends entirely on the basic properties of electricity and magnetism – Maxwell's great unifying theory in a nutshell!

The river analogy also gives us an idea of divergence, usually abbreviated to div. It is a measure of the excess of water flowing out of a small region surrounding our point compared with the amount flowing in. Assuming water is incompressible (very nearly true), these amounts are equal and the divergence is zero everywhere in the river. Unless, that is, we inject extra water at our point. The divergence then becomes

---

[†] The direction of the curl vector is conventionally taken as that in which a right-handed screw would move if it turned the same way. The term vorticity describes the variation of the vector in a small region immediately surrounding the point and does not imply movement unless, as in our river analogy, the vector is itself a rate of flow.

positive, and if we suck water out it becomes negative. The concept of divergence is needed to complete the description of how electric and magnetic fields behave. One of the two remaining Maxwell equations says that magnetic force is like the undisturbed river: it has zero divergence everywhere. The other says that electric force also has zero divergence where no electric charge is present but that an electric charge causes positive or negative divergence, depending on the sign of the charge. At a point in empty space there is no charge and the equations are simply:

$$\text{div } \mathbf{E} = 0 \qquad \text{div } \mathbf{H} = 0$$

Elegant and powerful, they state, in the compact language of vectors, that electric and magnetic forces obey an inverse square law.[11]

Amazingly, the four equations tell us everything about how electric and magnetic fields behave. In the presence of electric charges and conduction currents they acquire a few extra symbols but are still astonishingly simple. Their stark symmetry and power still hold scientists in thrall and they have aptly been called the Mona Lisa of physics.[12]

Like a beautiful tune, the equations seem so natural that they could have been plucked from the air, but their gestation was, in fact, quite complex. The concepts of curl and divergence were not new but it was Maxwell who brought them to the fore and gave them names.[13] He had originally set out his theory in ordinary mathematical notation but found that he could make things more compact by using some of the features of the system of quaternions invented by the Irish mathematician William Rowan Hamilton. 4nions, as Maxwell called them, were something like vectors but much more complicated, having both a scalar part and a vector part. They were not easy to get to grips with and only a few enthusiasts wanted anything to do with them. The foremost of these was Maxwell's friend P.G. Tait. Maxwell didn't share Tait's fiery enthusiasm for the full panoply of quaternions but tentatively adopted some of the ideas and notation – in effect paving the way for vector analysis. In his *Treatise* Maxwell played safe by writing out the equations twice – first in the standard way and then using a form of quaternion notation.

Heaviside had taken it from there. He thought that Hamilton's quaternions were elegant but largely useless, so he ditched the scalar components and worked out his own system, using plain vectors. Like many of his innovations, it met opposition. Vector analysis raises no eyebrows now but at first many orthodox mathematicians thought it difficult and peculiar. And there was fire from the other flank when P.G. Tait accused Heaviside of mutilating his beloved quaternions.

Stormy times ahead, but for the present Oliver calmly pressed on with developing the method of vectors and reworking Maxwell's theory. Unusually for him, he took great pains in paper after paper to explain everything clearly. Maxwell's original set of twenty equations contained six triples, which dealt separately with the x, y and z components of each vector, so once these were written in vector form the number of equations came down to eight. Maxwell had kept options open, remarking that the equations might readily be condensed but that 'to eliminate a quantity which expresses a useful idea would be a loss rather than a gain at this stage of our enquiry'. As was his way, Oliver experimented with the apparatus Maxwell had laid out. Even with the equations in vector form there remained an apparently impenetrable assortment of variables. For a while he remained as confused as anyone but then he saw a chink of light and got out his pruning knife again.

What he wanted was simplicity and symmetry, and he found them by concentrating on the electric and magnetic forces and cutting out the variables called potentials. The idea of potential came from mathematical astronomy, where it had helped to solve difficult problems.[14] Like gravitation, electrostatics worked by an inverse square law and it was natural for physicists interested in electricity to use the same methods – why reinvent the wheel? Electric potential was an indication of the energy held by a charged body by virtue of its position in the field; it was the *difference* in potential from point to point in the field that gave rise to forces. A rough analogy is river water held up by a dam: its high position relative to the water downstream gives it the potential of providing the energy, and the force, needed to drive a mill or a hydroelectric generator. Magnetic potential was much harder to understand. It was a vector rather than a scalar and represented a property of the magnetic field that was akin to momentum in mechanical systems. Mysterious indeed, but Maxwell thought it might be 'the fundamental quantity in the theory of electromagnetism' and gave it pride of place in his presentation.

Heaviside took a more down-to-earth view. To him, only the electric and magnetic forces were *real* quantities. The potentials were merely 'metaphysical' and he decided 'to murder the whole lot'. By allowing for the possibility of *magnetic* conduction currents – tantamount to assuming that north or south magnetic poles might exist without their partners – he worked out a system that had complete symmetry between electricity and magnetism. In deference to the fact that nobody had ever found a monopole he set the magnetic conduction current term to zero, but the symmetry was still there in principle. From this came the four famous 'Maxwell's equations', which summarised everything in terms of electric and magnetic forces.

Heaviside felt that he was being true to the *spirit* of the theory; that if Maxwell had lived he might well have taken the same course himself and, in any case, would have applauded the changes. He may have been right. Maxwell had been cut off while in full flow and one of the remarkable things about his electromagnetic theory was how little time he had spent on it. For him, electricity was only one of many interests. He had held professorships in Aberdeen and London before becoming founding Director of the Cavendish Laboratory in Cambridge and had made ground-breaking discoveries in thermodynamics, gas theory, colour vision and control theory. He somehow managed to do all this while conscientiously carrying out his duties as Laird of his inherited estate in Galloway. And nothing pleased him more than to find and encourage talented young scientists. He would, surely, have been delighted with Heaviside's work. When his favoured potentials were so rudely dismissed, he would probably have smiled indulgently like a father watching a boisterous child.

How do things stand today? By focusing on the electric and magnetic force fields, Heaviside has made the theory much easier to understand and more obviously relevant to everyday experience. His formulation has been used in most of the subsequent developments and applications of the theory. But not all: the wisdom of Maxwell's decision to keep options open still shines through. As the great American physicist Richard Feynman reported in 1964: 'In the general theory of quantum electrodynamics, one takes the vector and scalar potentials as the fundamental quantities.'[15] Honours even.

Heaviside had transformed Maxwell's theory and, in a sense, the process worked both ways. Some of Europe's finest physicists were also enthusiasts for Maxwell but had been struggling with the theory and they must have been astonished to find articles of such penetration and authority, written by someone they had never heard of who had no academic position and lived at an obscure address in Camden Town. From a nobody, whistling in the wind, Oliver was soon to be transformed into an esteemed sage, consulted and befriended by some of the outstanding scientists of the day. He would find true fellowship on his own terms. Good old Maxwell!

Chapter 6
Making waves
London, Liverpool, Dublin and Karlsruhe
1882–88

When Oliver began writing about Maxwell's theory in *The Electrician* in 1882, he probably felt like a voice crying in the wilderness. In fact, things were not so bleak: there were several others who had fallen under Maxwell's spell and each was trying to make his own way, interpreting, testing or developing the great man's ideas. By the end of the 1880s they had come together and, largely though their work, Maxwell's theory was beginning to take its place at the heart of electrical science and engineering, where it has remained ever since. Collectively, they had by then not only verified the theory by experiment but had recast it into the form in which it would eventually become standard fare for scientists and engineers. What is more, they had publicised their work effectively enough to overcome scepticism and entrenched ideas; they jolted scientific opinion on to a new track. The Maxwellians, as they came to be called, were not an organised group but simply a few outstanding individuals brought together by complementary talents and a common cause. Each took delight in the work of the others and in having appreciative, though often robust, criticism of his own. Along with Heaviside, the principal Maxwellians were Oliver Lodge, George Francis Fitzgerald and Heinrich Hertz.

As things turned out, it was yet another Maxwellian that Heaviside encountered first. Both of them had been intrigued by the problem of how electromagnetic energy moved. According to Maxwell, energy was *located* in space – at any instant, each part of space in the electromagnetic field contained a definite amount of energy, no matter whether the space was empty or occupied by matter. To most people this idea seemed strange enough, but it had an even stranger consequence that went beyond Maxwell's own work. When an electromagnetic field changed, some parts of the field would lose energy and others would gain. But energy could not vanish in one place and simultaneously appear in another, because this would involve action at a distance, the very concept that Faraday and Maxwell had set out to banish. So energy had to *flow*, like water in a river.

Oliver worked out how it flowed. The direction of energy flow was *at right angles* to both the electric and magnetic forces. For example, if the electric and magnetic forces were both in a horizontal plane, the electric force pointing east and the magnetic force north, energy would flow vertically upwards! And the density of the flow was given by the arithmetic product of these forces multiplied by the sine of the angle between them: it was greatest when the electric and magnetic forces were at right angles to each other. He expressed the rule compactly in vector form as:

$$\mathbf{W} = \mathbf{E} \times \mathbf{H}$$

where **W** is the energy flow vector and $\mathbf{E} \times \mathbf{H}$ is the vector product of the electric and magnetic forces.[1]

This was a great result but shortly after it came out in *The Electrician* Oliver found he had been scooped. John Henry Poynting, Professor of Physics at Birmingham University and a former student of Maxwell's, had published the same formula several months earlier in the Royal Society's journal. In fact, Heaviside had already given the formula in a paper published about the same time as Poynting's, but in words not symbols and only for a special case.[2] Heaviside's reaction to this disappointment gives us a little more insight into his character. On the one hand he gave generous credit to Poynting and never disputed priority, but on the other he never failed to mention his own part in the matter. For example, referring to the energy transfer formula in his later treatise *Electromagnetic Theory*, he wrote:[3]

> This remarkable formula was first discovered and interpreted by Prof. Poynting, and inde-
> pendently by myself a little later. It was this discovery that brought the continuity of energy
> into prominence.

But posterity prefers to keep things simple. The energy vector is known today as the Poynting vector – a title with a built-in aide-mémoire for students, as it *points* in the direction of flow.

Poynting gets the glory but it was Heaviside who did most to investigate and explain the extraordinary consequences of the new formula. With his background in telegraphy he was naturally interested in how energy travels along conducting wires, and what he found seems to defy belief, even today. No energy passed along the wires themselves (or through the earth, parallel to its surface, where that was used as a return conductor). The function of wires in a circuit was merely to act as a *guide* for a stream of energy which travelled alongside them through the surrounding space. Some of the energy from this stream did enter the conducting wires but as soon as it did so it changed direction and flowed *inwards*, at right angles to their surfaces. This was the portion of energy that was dissipated as heat.

What about the electric current – didn't that flow inside the wires? Yes it did, but the energy was borne not by the current itself but by the accompanying fields – the lines of magnetic force that encircled a current-bearing wire and those of electric force that spread out radially from it, like spokes. By the new formula, the direction of energy flow was at right angles to both these fields and so ran parallel to the wire.*

And what, exactly, *was* an electric current? Heaviside was a pragmatist: his job, as he saw it, was to work out the laws connecting things that could be measured and he didn't concern himself too much with speculation about nature's hidden mechanisms. What he knew was that a current always had a magnetic field around it, and that the

---

* Very nearly so, anyway. The lines of energy flow near the wire converged ever so slightly and, when they hit the wire, turned sharply inwards to be converted into heat.

magnetic force could be measured. *Magnetic force* was the physical manifestation of the current, and what went on in the wire was of secondary importance. For him, as for Faraday and Maxwell, electric and magnetic fields were the main thing: electric charges and currents were simply effects that the fields produced in conductors. He knew that moving charges produced magnetic effects, as currents did, but this did not *prove* that currents in wires were charges in motion. The discovery of the electron later settled this question – a current was indeed a drift of trillions of tiny charged particles in the wire – but a more fundamental question remained: what *is* an electric charge? Even today, quantum electrodynamics has only partly solved the mystery.[4]

The energy flow formula had another surprise up its sleeve. Since the energy entered a wire from outside, it followed that when an electric current started it flowed first in the outer part of the wire only and took time to penetrate to the middle. High-frequency alternating currents would scarcely penetrate the wire at all because soon after the current had begun at the surface of the wire its direction was reversed and the process would start over again. This became known as the 'skin effect' and it showed that signals moved along conducting wires in a way that was even further removed from the early telegraphers' water pipe analogy. At the frequencies inherent in telephony and high-speed telegraphy the current hardly entered the wires and, as Oliver put it, the signals just slipped over their surfaces 'like greased lightning'.

He had no doubt by now that in low resistance lines the signals were propagated through the space alongside the conductors in the form of waves and that these were Maxwell's electromagnetic waves. But his belief was based on intuition and mathematics; there was no physical evidence to back it up. All that people knew for certain about telephone and telegraph signals was what happened at the sending and receiving ends; to prove the existence of electromagnetic waves along the line one would need to detect them *in transit*. Meanwhile, rival theories to Maxwell's continued to flourish, especially on the continent of Europe. In his blunt way, Oliver described these theories as no more than 'absurd speculations', but only when someone supplied experimental proof of electromagnetic waves would the speculation cease.

Someone who was keen to supply the proof was Oliver Lodge. He was a man of tremendous energy, forceful and extrovert, who pursued his scientific career with a rare mixture of doggedness and passion. A year younger than Heaviside, he was the eldest son of a Staffordshire clay merchant and had been expected by the family to take over the business. After six bleak years at school he started work at fourteen but hated what he called 'the soul-destroying work of calling on Staffordshire potters and selling them clay'. On a visit to an aunt in London when he was sixteen he went to hear John Tyndall speak at the Royal Institution and was spellbound. From then on he knew exactly what he wanted to do in life. He endured the clay business for five more years but left as soon as he reached his majority, took night classes, did experiments in an improvised home laboratory, and worked his way to University College, London, where he was such a good student that the professor of physics made him his assistant. This gave Lodge a modest independence and he rented a room in, of all places, Camden Town, only a few streets away from Oliver Heaviside, though they didn't meet for another ten years. On the way to gaining a doctor of science degree in 1877, Lodge read Maxwell's *Treatise on Electricity and Magnetism*. It took him a while to find his way into Maxwell's thoughts but once there he was hooked.

What especially intrigued Lodge at first was the way that electrical force put materials (or empty space) under stress. A non-conducting material would absorb the stress like a spring, but in a conductor like iron or copper the material would give way and allow a current to flow. To help others grasp the idea, he built a clever teaching aid using string, pulleys, beads and screws on a simple wooden frame. By adjusting the screws it could be made to represent a perfect insulator, a perfect conductor, or anything in between. He was so pleased with it that he wrote to Maxwell about it and had a reply that was 'humorous and quite long'. Unfortunately, Maxwell's letter has not survived but we can be sure he approved. Maxwell had strong and progressive ideas on teaching, though he was not much good at it himself, and had summarised them neatly in a talk to the British Association for the Advancement of Science.[5]

> The human mind is seldom satisfied, and is certainly never exercising its highest functions, when it is doing the work of a calculating machine. ... There are, as I have said, some minds which can go on contemplating with satisfaction pure quantities represented to the eye by symbols, and to the mind in a form which none but mathematicians can conceive. There are others who feel more enjoyment in following geometrical forms, which they can draw on paper, or build up in the empty space before them. Others, again, are not content unless they can project their whole physical energies into the scene which they conjure up. They learn at what a rate the planets rush through space, and they experience a delightful feeling of exhilaration. They calculate the forces with which the heavenly bodies pull at one another, and they feel their own muscles straining with the effort. To such men, momentum, energy, mass are not mere abstract expressions of the results of scientific enquiry. They are words of power, which stir their souls like the memories of childhood.
>
> For the sake of persons of these different types, scientific truth should be presented in different forms, and should be regarded as equally scientific, whether it appears in the robust form and the vivid colouring of a physical illustration, or in the tenuity and paleness of a symbolic expression.

Maxwell could hardly have chosen better words to describe the two Olivers. Heaviside clearly fell into the second group: although he trained himself to be an effective manipulator and dispenser of symbols he was, above all, a master at building a mental picture of geometrical forms. Lodge, on the other hand, belonged to the group who felt their own muscles straining with the effort. And when muscles were not enough he conjured up mechanical models to inflict the stresses and bear the strains: after the beads and string device came a variety of others, some made up into pieces of apparatus, some imaginary, each designed to illustrate one or more aspects of the way that electricity and magnetism worked. He saw the need for mathematics and could, with difficulty, follow the work of others, but was never much of a symbols man. The comments of one fellow-physicist on a book he wrote a few years later, *Modern Views of Electricity*, give a good idea of his style.[6]

> Here is a book intended to expound the modern theories of electricity and to expound a new theory. In it there are nothing but strings which move around pulleys, which roll around drums, which go through pearl beads, which carry weights; and tubes which pump water while others swell and contract; toothed wheels which are geared to one another and engage hooks. We thought we were entering the tranquil and neatly ordered abode of reason, but we find ourselves in a factory.

He may not have impressed this reviewer, but his 'factory' approach brought ideas home to people in a compelling and vivid way. Lodge was a confident and effective speaker and teacher, and quickly made up for the delayed start to his career, publishing papers about his models in the *Philosophical Magazine* and making his mark at meetings of the British Association for the Advancement of Science. In 1881, aged thirty, he applied for the post of Professor of Physics at the new University College in Liverpool and got the job.

While still a student he had been captivated by Maxwell's idea of electromagnetic waves and by the possibility of proving their existence by generating and detecting them. Looking back from our distant viewpoint, it seems strange that physicists everywhere were not striving for this great prize, but, in fact, Lodge was one of the first people to give the matter serious and sustained attention. The main reason the task had been neglected was that it was so different from anything else that had been done by experiment that nobody knew how to set about it: Maxwell's theory predicted the waves but didn't tell you how to produce and detect them in a laboratory. One might have expected Maxwell himself and his students at the Cavendish to have had a go, but in Maxwell's time the new laboratory was under critical scrutiny from a sceptical public and needed to establish its reputation with some early successes; experiments on electromagnetic waves would have been far too hit-or-miss for this purpose. If the theory was correct then it would not be difficult to generate waves. In fact, one couldn't help but produce waves in ordinary operations with an electrical circuit, like switching a current on or off: the problem lay in detecting them. Conversely, light waves were easy to detect and, according to Maxwell's theory, electromagnetic, but the problem lay in producing them by any known electromagnetic mechanism because the frequencies required were enormous – more than a million million cycles per second. With the blithe confidence of youth Lodge decided to try.

His first idea was to pass a current across the contact point between a rapidly spinning carbon disc and another piece of carbon held against the disc by a spring. The idea came from David Hughes' new invention, the carbon microphone: Lodge's device was, roughly speaking, a microphone with its sound board replaced by the spinning disc. He reasoned that as the disc spun and the sprung contact piece bumped over the tiny humps and dips on its surface, the resistance of the contact would fluctuate very fast and so would the current. If the fluctuations were fast enough the circuit would radiate light waves and one would see the glow. A bold idea but hopelessly naïve; it didn't get anywhere near the required frequency. His next idea was that a current sent through a vacuum tube might fluctuate at the right frequency, but that came to nothing.

Then he had the notion of passing a 'chopped' current through a series of coils. A current would be switched rapidly on and off by mechanical or electromagnetic means, giving several hundred cycles per second, and the effect of each successive coil would be to double the frequency. By about the fortieth coil the frequency would be high enough to produce light! This was rather like the idea that a piece of paper folded on itself fifty times will reach to the moon. Repeatedly doubling paper thickness by folding doesn't work because in practice only the first few folds are achievable, and in Lodge's plan frequency doubling would only have happened in the first three or four coils: after that, the edges of the current pulses would no longer have been sharp enough to bring it about.

Lodge was saved some unprofitable labour when a friend pointed out the futility of the scheme. The friend was George Francis Fitzgerald. They had met two years earlier when the British Association for the Advancement of Science held its 1878 meeting in Fitzgerald's home city, Dublin. The two men were the same age and quickly developed an easy comradeship. Both were admirers of Maxwell's electromagnetic theory and each was delighted to find a fellow enthusiast with whom to try out his own ideas on how to interpret and develop it. They stayed in close touch, writing frequently and meeting when they could.

Fitzgerald's path in life was smoother than that of Lodge or Heaviside. He was born into one of Ireland's patrician Protestant families and spent his first five years in the milieu of Trinity College, Dublin, where his father, William, was Professor of Moral Philosophy. It was not unusual in those days to combine an academic career with one in the Church and William did this in style, becoming Bishop of Cork in 1857 and Bishop of Killaloe five years later. After his mother died when he was eight, George was taught at home by private tutors, along with his two brothers. Although uncharacteristically slow at languages at first, he mastered his other lessons without apparent effort and easily outshone his fellow students when he entered Trinity College at sixteen. After taking his degree in mathematics and experimental science with top honours in 1871, he set out to win one of the esteemed Trinity fellowships. Vacancies occurred only rarely and were fiercely competed for in a four-day examination which included classics and philosophy. When the first vacancy came up, Fitzgerald, for the first time in his life, lost. But this was only a temporary setback. He won on his second attempt, was awarded his fellowship in 1877 and went on to become Professor of Natural and Experimental Philosophy four years later.

A man of striking appearance, athletic and distinguished, Fitzgerald had natural gifts in such abundance that everything seemed to come easily to him. Perhaps for this reason, he lacked the stern mental discipline that comes from sustained toil and rarely pursued ideas to their limits. Heaviside later said of his friend: 'He had, undoubtedly, the quickest and most original brain of anybody ... He saw too many openings. His brain was too fertile and inventive.' In his unpretentious way, Fitzgerald claimed he was simply too lazy to think out the consequences of his ideas, but he dispensed them generously and seemed to take as much pleasure from helping others as from making his own discoveries. Together with this philanthropy went an infectious enthusiasm; he was the most stimulating of colleagues and had an influence on late nineteenth-century physics way beyond the range of his own published work.

At the time he met Lodge, Fitzgerald was absorbed in magneto-optics: trying to explain how the observed phenomena of light such as reflection and refraction could result from purely electromagnetic effects. Characteristically, his work in this field did not itself solve the problems of the day but had a great influence on later developments by Hendrik Anton Lorentz and Joseph Larmor. Also typical was the way he responded to comments on his paper by Maxwell himself, who had refereed it for the Royal Society shortly before he died. Maxwell had pointed out the need to make the physical meaning of the equations clearer. Fitzgerald acknowledged the fault but had some difficulty putting things right because he hadn't kept a copy and couldn't remember exactly what he had written. In some embarrassment, he confessed to the Society's secretary, George Gabriel Stokes, that he had 'got very tired of the paper long before he had finished writing it'. Nevertheless, he did put things right and later gave encouragement and help to Lorentz and Larmor.

Before talking to Lodge, Fitzgerald had shown little interest in the possibility of producing and detecting electromagnetic waves but he then took up the topic with, for him, rare tenacity. After putting Lodge straight on the impracticability of generating light by known electromagnetic means, he realised that waves like light but of much longer wavelength might be radiated from an electrical circuit in which the current was made to oscillate at a rapid but achievable frequency. Maxwell himself had not done the calculations to show how much energy would be radiated from an oscillating circuit and Fitzgerald took up the task.

Almost at once, he stumbled. Maxwell had stated in his *Treatise* that his own theory gave results that were 'mathematically identical' to those of the rival action-at-a-distance theories and, taking this literally, Fitzgerald followed a false trail which led to the conclusion that the waves produced by an electrical circuit would simply swing back and forth and would never spread out into space. In fact, Maxwell had meant his statement to apply only when currents were steady and no waves would be produced anyway, but Fitzgerald was thoroughly misled and took two years to realise his mistake. He found the error himself, described his early ideas as 'more or less idiotic', and discovered the true path. An oscillating current would indeed send waves into space, although only a fraction of the total energy of the field would escape this way – the rest would be held captive, as it were, by the circuit.

This was a great step: Fitzgerald had separated what came to be called the near field, which was restrained by the circuit, from the far field, which radiated energy out into space. Going further, he designed an antenna – in the form of a circular wire loop – and worked out a formula for the amount of energy that it would radiate per second when carrying an alternating current. Strikingly, the formula showed that the amount of radiation was proportional to the *fourth power* of the frequency: by making the current oscillate ten times faster you would get ten thousand times as much radiation. At frequencies of a few thousand cycles per second the radiation would be weak and the wavelength huge but at, for example, 30 million cycles per second the waves could well be strong enough to be detectable and their wavelength would be only ten metres, small enough to be measurable inside a laboratory. He became convinced that the means of producing frequencies of this order already existed: all you had to do was to discharge a capacitor, such as a Leyden jar, through a suitable circuit and the waves would be radiated. Fitzgerald was not himself much of a hand at experiments but was always inviting others to take up his ideas and this time he went as far as to publish the invitation. At the British Association meeting at Southport in 1883 he presented his suggestions, which were summarised for publication in one of the shortest papers ever written; here it is, complete with title.[7]

On a Method of Producing Electromagnetic Disturbances of Comparatively Short Wavelengths

This is by utilising the alternating currents produced when an accumulator[†] is discharged through a small resistance. It would be possible to produce waves of as little as 10 metres wavelength, or even less.

He wasn't giving much away here! The paper was more of a mission statement than a plan: presumably he was writing for people who already knew that to get a short wavelength you needed to keep both the inductance and capacitance of the circuit low. And there was no mention of the important matter of how to detect the waves, but Fitzgerald had two good ideas here. One was to use a well-known property of all oscillatory waves: reflect them back directly towards the source and the forward and backward components combine to form *standing* waves. Since standing waves appear simply to oscillate in the same place they are much easier to detect and study. The other idea, which later became familiar to anyone who operated a wireless set, was to use a *tuned* receiver, one with a circuit that resonated at the same frequency as the waves. Fitzgerald was on the right track with both these ideas but a vital component was still missing. As he explained to a colleague: 'The great difficulty is something to *feel* these rapidly alternating currents with.' No known instruments seemed to be up to the job, but a few years later somebody did indeed find a way to 'feel' the currents. It was simple, effective and, tantalisingly, had been available all along.

We shall soon come to this, but, meanwhile, a drama was being played out. Practical-minded Lodge was just the man, one would have thought, to take up the search for an effective detector but he always deferred to Fitzgerald in theoretical matters and had been thoroughly discouraged by his friend's initial conclusion that no electrical circuit could produce waves. By the time of Fitzgerald's spectacular conversion from sceptic to enthusiast, Lodge had taken up his post at Liverpool and was busy getting the new department of physics established. Teaching took up five hours a day and that was only part of the job. For several years there was no time even to think of experiments on electromagnetic waves. By early 1888 things had eased a little and he accepted an invitation from the Society of Arts in London to give two lectures on a topic of great public concern – how to protect buildings against lightning. It was his reputation as a powerful speaker that had led to the invitation. He was no expert on lightning and to get a feel for the subject decided to try some experiments with Leyden jars. The sparks produced when a jar discharged would be a proxy for the real thing – they looked like lightning and would, surely, have similar properties.

Following up this assumption led him quickly to strong views about lightning conductors which conflicted head-on with those of the man who considered himself the country's foremost expert on the topic, none other than Heaviside's arch-enemy, William Preece. For once, Preece was right. Lodge had been misled by the apparent

---

[†] 'Accumulator' did not have its present-day meaning of battery. Fitzgerald meant a capacitor.

similarity of two different phenomena. In his Leyden jar experiments the discharge current oscillated very rapidly – a million cycles per second or more – and he assumed that lightning flashes did the same. In fact, as later experimenters found out, lightning does not generally behave in this way. The debate came to be fought out in public, like a heavyweight boxing match. Although, as it turned out, Lodge had built his case on sand, he used the occasion to expose Preece's complacent ignorance on scientific matters and, in doing so, helped Heaviside in his own quarrel with Preece. We'll return to this in the next two chapters.

Lodge's experiments were important in another way and it had nothing to do with lightning. Maxwell used to say 'I never try to dissuade a man from trying an experiment; if he does not find what he wants, he may find out something else.' And this is exactly what happened. When Lodge discharged his jar, sparks jumped between the free ends of two wires connected to it. Nothing remarkable so far, but he found by chance that the sparking could be made weaker or stronger by changing the length of the wires. This was interesting but puzzling. Only when a junior colleague prompted him did Lodge realise that he had stumbled upon evidence of Maxwell's electromagnetic waves. Eureka! He reasoned that waves from the discharge must be streaming through the space alongside the wires and being reflected back from the free ends; the sparking was, as he put it, a kind of 'recoil kick' from the waves being fired back. What he had indirectly detected were the resulting standing waves. His second lecture on lightning protection was coming up and it was in that unlikely context in February 1888 that he presented a first brief sketch of the long-awaited physical confirmation of Maxwell's theory. Lodge knew that to get full recognition of his achievement he would have to carry out much fuller and more rigorous experiments but there was plenty of time to do this before the British Association for the Advancement of Science's meeting at Bath in September.

Meanwhile, he and Heaviside discovered one another. Just at the time of his lucky strike, Lodge was looking through the *Philosophical Magazine* when his eye was caught by the title 'On Electromagnetic Waves'. It was the first in a series of papers on the topic by a Mr O. Heaviside and was like nothing he had seen before. The writing was idiosyncratic but brimming with confidence and authority. Lodge looked up the author's earlier work and it was a revelation. Heaviside had not only interpreted Maxwell's ideas but also extended them. Here was a deep and detailed mathematical treatment of the very subject Lodge had been tackling with his mechanical models and it gave exactly the theoretical underpinning needed for his experiments on waves along wires. In his second talk to the Society of Arts, Lodge acclaimed Heaviside's work, though with a warning that some may find the gratuitous observations on sundry topics like archbishops to be in poor taste:[8]

> I must take this opportunity to remark what a singular insight into the intricacies of the subject, and what a masterly grasp of a most difficult theory are to be found in the eccentric, and sometimes repellent, writings of Mr. Oliver Heaviside.

What would Heaviside make of this? He was overjoyed. What was 'repellent' when set alongside 'masterly'? As we'll see in the next chapter, Heaviside was by now embroiled in a bitter dispute with Preece and the other 'practical' men who held sway in the electrical engineering establishment. His work was being spurned, even suppressed, and he desperately needed publicity to gain support among the wider scientific community. He later told Lodge: 'I looked upon your 2nd lecture when I read it as a sort of special providence!'[9]

Heaviside lost no time in contacting his new champion. Lodge sent him a full copy of the lecture text and continued to sing his praises. Compared with Heaviside, he wrote to *The Electrician*, 'no person now alive has anything like the intimate familiarity with the properties of alternating currents'. Heaviside, in turn, congratulated Lodge on having added substantially to 'our knowledge of sudden discharges'. Both were gratified by the approbation of the other but it was more than mutual admiration, more even than mutual advantage, that pulled them together. One effect of Maxwell's legacy was to forge a tremendously strong bond among the men who caught the spirit of his work on electromagnetism and took it forward.

Meanwhile, Fitzgerald had, like Heaviside, been working on the theory behind the propagation of electromagnetic energy and waves, but he had become enmeshed in a thicket of electric and magnetic potentials. After several years of frustration, he was beginning to think that there must be a better way to make progress when – probably prompted by Lodge – he looked to see what Heaviside had done. Genius was laid bare: Heaviside, too, had at first been tied down by the potentials but had cut the Gordian knot by discarding them and recasting the theory using only the electric and magnetic forces. Fitzgerald had his mathematical roots in the refined Dublin school, and Heaviside's earthy style came as something of a shock, but, like a true intellectual, he was delighted to be made to look at things in a different way. Here was somebody with a mathematical imagination on a par with his own, who could help to bring order to his own teeming ideas. Fitzgerald and Heaviside came together just as Lodge and Heaviside had done.

The benefits ran the other way, too. Not only was Fitzgerald a congenial and stimulating colleague, he commanded tremendous respect and did more even than Lodge in getting Heaviside's work recognised. He was probably the first person to follow and appreciate Heaviside's mathematics properly and it was largely through his advocacy that other academic physicists began to take serious notice of Heaviside's papers. Lodge, too, was better able to appreciate Heaviside's more mathematically abstruse writings now that he had Fitzgerald to explain them and this brought the three of them even closer together. The three men, in their mid-thirties and at the height of their powers, formed the core of the expanding Maxwellian club. They were about to be joined by a fourth – a younger man from a surprising source.

Lodge prepared keenly for the September 1888 meeting of the British Association, where he expected to make a big splash with an account of his experiments. By now he had incontrovertible evidence of waves along wires, carefully observed and logged, and planned to present them together with suggestions for the next stage – detecting waves in free space. In high spirits, he left for a summer holiday in the Alps, taking with him some journals he had not had time to read. One of them was the July issue of the German journal *Annalen der Physik und Chemie*. He opened it on the train just after leaving Liverpool and could scarcely believe his eyes. A Dr Hertz of the Technische Hochschule at Karlsruhe had already produced and detected electromagnetic waves in free space, as well as along wires. And he had gone further, showing that the waves could be reflected, refracted and polarised, just like light.

Poor Lodge. It was as though he had gone in for a piano competition, having just mastered his Grade 8 piece, and had entered the competition hall with high hopes only to have them shattered by hearing a fellow competitor playing Beethoven's Appassionata Sonata with the assurance of Daniel Barenboim. But the disappointment was soon overtaken by admiration of Hertz's beautiful experiments and genuine pleasure at the results. The report on his own findings would now play only a small part at the British Association meeting at Bath. But there was a much bigger story to tell. Speculation was over; Maxwell's theory had triumphed. It now pointed the way to all future developments in electrical physics and engineering; action at a distance was dead. And a question in the minds of many delegates would be: who is Heinrich Hertz?

Like Fitzgerald, Hertz was born into a comfortably-off family with a strong intellectual tradition. His father, who became a senator of Hamburg after a successful career as a barrister, came from a line of Jewish merchants, prominent citizens of that city. They had wide interests and the money and leisure to pursue them. Young Heinrich acquired his father's gift for languages and his paternal grandfather's love of science. The old man had built his own laboratory and, when he died, Heinrich, still a boy, inherited the apparatus. Heinrich's mother was not Jewish, but the daughter of a doctor who was descended from generations of Lutheran preachers in South Germany.

With this eclectic background, the boy developed a free-ranging and independent outlook coupled with a remarkable degree of self-discipline. At school he was faced with choosing between the old *klassiker* path, based on Greek and Latin, and the new *techniker* one that was sweeping the country. Rather than follow either course to the exclusion of the other, he devised his own way to combine them, changing schools several times and, at one stage, studying at home. He enjoyed making apparatus for his home laboratory and decided to try his hand at engineering. Not one to do things by halves, he joined an engineering firm at eighteen for a year's practical experience before studying at Dresden and Munich, with a year's military service in between. At Munich, still only twenty, he came at last to see that his true vocation was in mathematical and experimental physics and set his course in that direction. A remarkable feature of this unusual career as a student is how smoothly he seems to have fitted in at every stage. It was not unusual in those days for German students to change schools and universities in the course of their studies but Hertz showed such prodigious talent at everything he did that he seems to have been given exceptional freedom: he could come and go as he pleased. And now that he was set on a scientific career there was only one place to go. In 1878 he left Munich for the University of Berlin, as a later writer put it, 'to acquire the stride of the giants'.[10]

One of the 'giants' was the German counterpart of William Thomson, Hermann Helmholtz, a powerful and immensely respected scientist who was shortly to be ennobled, entitling him to add a 'von' to his name. Helmholtz had a vast range of interests in physiology and physics but the topic that absorbed most of his attention at this time was electromagnetism. The dominant theories in Germany were those of Weber and Neumann, which both rested on the assumption that electrical charges and currents somehow acted on one another at a distance, with the intervening space playing only a passive role. The contrary idea that space itself was an agent for transmitting electric and magnetic forces seemed weird, and few people outside Britain took Maxwell's theory seriously at first. But Helmholtz was one who did and by the time Hertz arrived he had succeeded in persuading Berlin colleagues first that the three theories were all legitimate contenders and, second, that it was important to find out by experiment which was correct.

As a step towards this goal, the University set a prize competition to investigate by experiment 'the inertia of electricity in motion' – in essence, Heaviside's 'extra current'. Inspired by Helmholtz's blend of erudition and enthusiasm, Hertz collected some of his apparatus from home and rattled off a series of experiments. Despite having to write up the results during another stint of military service, he won the gold medal. The venture not only established Hertz as a rising star, it awakened a fascination with the ways of electricity almost as intense as that which had gripped Heaviside a few years earlier. He gained his doctorate, together with the rare award *magna cum laude*, by investigating such subjects as electromagnetic induction in rotating spheres and became Helmholtz's right-hand man, acting as his demonstrator and fielding students' tricky questions on electrical theory. Prompted by Helmholtz, he started trying to devise an experimental test of Maxwell's theory. The idea at first was to detect not waves but rather the fleeting currents which the theory predicted would occur in insulators whenever the electric and magnetic fields changed – Maxwell called them displacement currents. The only way they knew to prove that these tiny currents existed would be to demonstrate their magnetic effect and Hertz doubted whether such a small magnetic force would register on even the most sensitive instrument. Still, he tried, and although he found nothing, the work further honed his experimental skill so that when luck presented him with the slenderest opportunity a few years later he was ready to exploit it.

The next step was to gain some teaching experience to qualify for a professorship, so in 1883 Hertz took a job as an unpaid lecturer – common practice in those days – at the University of Kiel. He worked there for two years, taught well, and, as the university had no suitable laboratory facilities, turned his research efforts to theory. A diary entry records:[11]

> Hard at Maxwellian electro-magnetics in the evening. Nothing but electro-magnetics. Hit upon the solution of the electromagnetic problem this morning.

We don't know what particular problem this was but, as we shall see, he was coming to views similar to Heaviside's, though from quite a different angle.

In 1885, he applied for the professorship of experimental physics at the Technische Hochschule in Karlsruhe, and got the job. It was a happy time; he fell in love with the daughter of one of the lecturers at nearby Karlsruhe University and they were married within a year. By then he was back at experimental work in the splendidly equipped Hochschule laboratory, still trying to detect the slightest sign of Maxwell's displacement currents in insulators. In the end he succeeded, but in the course of these attempts found a far more convincing way to verify Maxwell's theory.

Playing with some of the stock apparatus, he found that when he switched on a primary spark-generating circuit not only did sparking occur across the terminals of a secondary circuit a little distance away (an expected result from electromagnetic induction) but side sparks also appeared from a neighbouring wire. What was going on? At this stage he didn't really know what he was looking for but, prompted by intuition, he felt his way towards a great discovery. In the course of trying many variations in the arrangement of the circuits he found that the intensity of the sparks in his secondary circuit varied as he moved it along near the wire; there was even a neutral point where the sparking stopped. Here was a standing electromagnetic wave, and it could only have resulted from the combined effect of forward and reflected *travelling waves* – Maxwell's waves.

Hertz had shown that electromagnetic waves travelled alongside wires – two years before Lodge made the same discovery. Both had taken advantage of chance observations but Hertz had gone further in this direction than Lodge. He had found the magic instrument that others had sought in vain – something to 'feel' the waves with – and it was nothing more than a loop of wire with a spark gap. If the detector loop was the right size and shape, the waves would set it resonating and generate enough electromotive force in the wire to make sparks jump across the gap. Naturally, things were not quite that simple. Even for Hertz, the most gifted of experimenters, results came only after a great deal of trial and error and were the fruit of patience and determination as much as skill. For example, at every stage he tried different configurations of primary (transmitter) and secondary (detector) circuits to find which gave the clearest and most consistent results.

Waves along wires were exciting but Hertz realised that the clinching test of Maxwell's theory would be to see whether they also travelled through space (or air) with no wire to guide them. If they did, his loop detector, with its spark gap, would be just the thing to find them. Making sure his transmitter was well away from stray wires, he made the detector as sensitive as he could by setting its spark gap jaws close together and looked for sparks. Tiny flashes appeared – faint but unmistakeable. A good check to make sure that the sparking truly resulted from electromagnetic waves would be to produce a standing wave, so he set up a metal plate to act as a reflector and looked to see if the brightness of the sparks varied as he moved the detector from the source towards the reflector. Indeed they did: there were nodes in the standing wave, where there was no sparking at all, and antinodes where the sparking was strongest. He had produced and detected electromagnetic waves in space.

This was a historic moment in the history of physics. Hertz had triumphed, and so had Maxwell, whose strange prediction had baffled most of his contemporaries a quarter of a century earlier. But this was just the beginning of a brilliant series of experiments in which Hertz, tentatively at first, then with progressively greater confidence, examined every aspect of the new waves.

By varying the oscillating frequency of his transmitting circuit and tuning the receiver loop to match, he managed to produce and detect a range of waves with wavelengths as long as 9 metres and as short as 50 centimetres. The new waves passed through thick wooden doors but were reflected by any metal surface, like light by a mirror, and he showed that they could be refracted in the same way that light is through glass: for this purpose he built a prism of solid pitch that stood as tall as a man and weighed about a ton. He knew the frequency of his transmitting circuits by calculation from their values of inductance and capacitance, and this allowed him to measure the speed of the waves in air from the simple formula: speed equals wavelength times frequency. It turned out to be the speed of light: in a sense they *were* light waves, but of much lower frequency and longer wavelength than visible light. We use them today as radio waves.

Having shown that the waves could be reflected and refracted, Hertz wanted to complete the picture by showing that they could also, like light, be polarised. This was tantamount to proving that the waves resembled light by being transverse, like ripples travelling along a rope, rather than longitudinal, like sound waves. In polarised light, the transverse wave vibrations are all lined up, rather than being higgledy-piggledy, and Hertz knew that this could be achieved by passing light through a crystalline substance which filtered out vibrations in all directions but one. How could he make a giant crystal that would do the same for his much longer waves? The solution, like many of Hertz's ingenious improvisations, seems simple in hindsight. He built a wooden frame, two metres tall, with parallel copper wires, three centimetres apart, stretched across it like piano strings. When he sent waves through the grating, any electrical vibrations parallel to the wires were filtered out and only those at right angles to the wires remained – the waves were indeed polarised and behaved just like polarised light.

In 1888 Hertz presented his findings in possibly the greatest series of experimental papers ever written – it was the second of these that Lodge had read on the train out of Liverpool. They were written in a matter-of-fact style with no fanfare and made little impact at first in Germany. Given Helmholtz's expressed enthusiasm for putting the rival theories of electromagnetism to the test one might have expected great excitement at Hertz's achievement, but German physicists were slow to grasp its full significance. They had been reared on the action-at-a-distance theories of Weber and Neumann, and even Helmholtz had interpreted Maxwell's theory in a way that held on to some of the action at a distance notions and was, according to the British Maxwellians, plain wrong. It failed to explain the flow of electromagnetic energy in space, and without energy flow there could be no waves, no matter how you jiggled the mathematics. As usual, Heaviside put it bluntly:[12]

> Helmholtz's theory seems to me as if he had read Maxwell all at once, then gone to bed and had a bad dream about it, and then put it down on paper independently; his theory being Maxwell's run mad.

But Hertz was no rebel; he had immense respect, even reverence, for Helmholtz and did not realise at first what a catastrophic blow he had struck. He knew his results were significant but it took him a while to recognise that they had smashed his mentor's ideas to pieces. Still less did he see that by demolishing action at a distance and vindicating Maxwell he had set a new direction for all future developments in electrical science and technology. He was content to let the results speak for themselves and for others to judge their importance.

The British Maxwellians felt no such restraint. They had long embraced what they felt sure was the true interpretation of Maxwell's theory, unfettered by any action-at-a-distance notions, and had made it their own. They were utterly convinced of its correctness. All they lacked was strong physical proof to win over sceptics, and this Hertz had handsomely supplied. He was, quite simply, a hero. They showered him with praise, welcomed him into their ranks with joy and set about promoting his work with tremendous enthusiasm. A later writer reported, with no more than slight exaggeration, that news of Hertz's discoveries reached Germany by way of England.[13]

Hertz completed the quartet at the core of the Maxwellian movement. A modest and genial man, he was surprised at the almost rapturous acclaim of his work in Britain but was delighted to make the acquaintance of Lodge, Fitzgerald and Heaviside. He entered whole-heartedly into lively correspondence with each of them and they became a close and mutually supportive group. They exchanged off-prints of papers as routine and, like good friends, were candid in criticism of each other's work. For example, Hertz joined Fitzgerald in telling Heaviside that if he wanted people to read his papers he would have to make them easier to understand. Oliver took no notice, of course.

Everything was set for the informal but highly effective Maxwellian group to establish its influence and bring about a transformation in the accepted theory and practice of electrical science. We shall return to the group's glory days, but first need to catch up with some dramatic, and traumatic, events in Heaviside's life.

Chapter 7
Into battle
London 1886–88

By the time he struck up an alliance with Lodge, Fitzgerald and Hertz in 1888, Oliver Heaviside had been living much the same life for fourteen years. The daily routine rarely varied – reading, exercise and experiments during the day, followed by theoretical work into the night. His only professional colleague was his brother Arthur, for whom he sometimes acted as unofficial and unpaid scientific adviser. The informal arrangement suited them both. Arthur had become an important man in the Post Office – Senior Engineer of the Northern Division – and valued Oliver's views on technical matters. In return, he sent Oliver equipment for experiments and kept him abreast of news in the telegraph and telephone industry.

Contact with Arthur gave Oliver a small window on the world but by the mid-1880s the loneliness had begun to tell. The early feeling of self-sufficiency – that his discoveries were all the 'meat, drink and company' he needed – was wearing thin. It was no longer enough simply to believe that his work would be recognised and appreciated one day. He needed his voice to be heard, to have some *influence* on what was going on in the world.

The prolonged obscurity was partly of his own making. For most readers of *The Electrician*, his papers might as well have been written in Egyptian hieroglyphics, and he made little effort to help them. Fitzgerald later described the task confronting anyone who made a serious attempt to understand Heaviside's work.[1]

> Oliver Heaviside has the faults of extreme concentration of thought and a peculiar facility for coining technical terms and expressions that are extremely puzzling to a reader of his Papers … In his most deliberate attempts at being elementary, he jumps deep double fences and introduces short-cut expressions that are woeful stumbling blocks to the slow-paced mind of the average man.

A fair summary of how things seemed at the time, but there was more to Heaviside's papers than readers of the 1880s, even Fitzgerald, could have imagined. They would be astonished to discover that some of the strangest-seeming parts of Heaviside's work, for example his expansive introduction to Maxwell's theory, present no difficulty to today's readers. Their 'terms and expressions' are no longer puzzling because they are the ones that have now come into general use. The arguments now seem natural and his system of vector analysis that baffled his contemporaries is part of the standard language of physics and engineering, as are many of the words he introduced, such as inductance, impedance and attenuation. Terminology was a particular fad. He wrote a five-paper series called 'Notes on Nomenclature' and justified this apparent obsession by saying: 'consider what frightful names might have been given to the electrical units by the Germans'. He may not have been wholly in jest: before settling on the admirable word *Fernsprecher* for telephone, the German authorities had contemplated alternatives such as *Doppelstahlblechzungensprecher*.

Heaviside was like an entertainer performing a new show on his own to a nearly empty theatre. It was a good show that would eventually pull in appreciative audiences but for the present almost nobody paid it any attention. The prevailing attitude towards its topic, the theory of electrical communications, was 'who needs it?' The great irony was that the people who disparaged theory the loudest were the very ones who needed it most – the leading electrical engineers. Their case was simple and, on the face of it, compelling. Telegraph networks, which spanned the world and carried thousands of messages each day, had been designed, built and run by practical men; theory had played little part in the enterprise. The new wonder of the age, the telephone, similarly owed little to theory. What counted was intuitive flair, a cool head and, above all, experience. Electrical engineering was something you developed a feel for, like riding a horse: mathematics hardly came into it.

What the self-proclaimed 'practical men' had failed to see was that the technology of electrical communications had run into the sand. Flushed by the runaway success of the telegraph, they saw themselves as masters of electricity and had no time for dilettante theoreticians. When faced with the puzzling behaviour of electrical circuits carrying high-frequency oscillating currents they came up with explanations that were often not only wrong but absurd.

Oliver's frustration grew and two incidents in the mid-1880s brought it to a head. The first, in 1886, concerned some highly publicised experiments by David Hughes, the brilliant and respected inventor who had just become President of the Society of Telegraph Engineers and Electricians. Born in Britain in 1831, Hughes grew up in America after his parents emigrated and as a young man had a spectacular success with a new type of telegraph printer.[2] He returned to London but his printer failed to break Wheatstone's grip on the British market so he took it to Paris and made another fortune from European sales. Back in London in 1879, he narrowly missed gaining a place in history as the first man to produce and detect electromagnetic waves. When a transmitter in his house was switched on he could hear crackling sounds in a portable acoustic receiver as he walked up and down the street outside. What he detected may well have been waves but he knew nothing of theory and when advised by William Preece and others that what he had detected was nothing more than ordinary electromagnetic induction he dropped the project. His golden touch with inventions did not desert him; among his later creations was the carbon microphone.

Hughes chose his Presidential Address to present the results of some experiments in which he examined the behaviour of metal conductors of various shapes, sizes and composition when subjected to a rapidly varying voltage, and compared it with how they behaved when the voltage was steady. His aims were good ones. He wanted to bring home to rank and file members that the usual formula for Ohm's law did not apply to circuits with rapidly fluctuating currents, and to report authoritatively on the different properties of each type of conductor in both steady and fluctuating conditions. But he got hopelessly muddled, both in the design of the experiments and in interpreting the results, and managed to confuse himself along with everyone else. He muddied the waters even more by taking the trouble to read the relevant part of Maxwell's *Treatise* and drawing faulty inferences from it. Although his circuits had actually behaved in a way that was consistent with Maxwell's theory, Hughes concluded otherwise and went as far as to suggest that Maxwell might have been wrong. Such words from a man of great prestige and high office carried weight and word spread of Hughes' new discovery.

Oliver tore his hair. He had been writing articles on the behaviour of electrical circuits for sixteen years and had set out in detail the theory underlying the very aspect that Hughes was investigating – rapidly oscillating currents. And yet here was the chief representative of the electrical profession propounding arrant nonsense that displayed ignorance of even the most basic theory. It was a both a resounding slap in the face and a spur to action.

Unlike Preece, Hughes was not an aggressive denigrator of theory. He had by no means intended to demean theoreticians; in fact, he thought he was giving them interesting new material to work on. Had Heaviside written privately to Hughes, tactfully pointing out his errors and offering cooperation in further experiments, he would probably have had a friendly reply. But we have seen enough of Oliver to know that any kind of diplomacy was beyond him; the only tactic he knew was to attack.

Reading through the published version of Hughes' talk, he hardly knew where to begin his assault. Terms such as 'resistance' and 'inductive capacity' were used to mean different things in different parts of the paper, the analysis was hopelessly inadequate and the design of the experiment seems to have been incapable of yielding sensible results in any case. But it didn't take him long to identify the main source of Hughes' confusion and muddled results. It was what came to be called the skin effect. When a current first starts to flow it always begins on the outside of a conductor and takes time to spread to the centre. A rapidly oscillating current never gets time to penetrate far below the skin and so the effective cross-section of the conductor is reduced, thereby increasing its resistance and reducing its inductance. All this followed from Heaviside's pioneering work on energy flow which had been well documented in his papers, but Hughes clearly had no notion of it.

Mustering what politeness he could, Oliver wrote in *The Electrician*:[3]

> Owing to the mention of discoveries, apparently of the most revolutionary kind, I took great pains in translating Prof. Hughes' language into my own, trying to imagine that I had made the same experiments in the same manner (which could not have happened), and then asking what are their interpretations? The discoveries I looked for vanished for the most part into thin air. They become well known facts when put into common language.

Hughes was an old hand at battles of words, having tangled with such rivals as Thomas Edison over patent rights, and was quick to respond.[4]

> ...if we assume with Mr Heaviside that all my results are contained in some mathematical formula of which I am unaware, then the mathematicians ought to be grateful for the experimental proof I have furnished them.

Oliver would, indeed, have been grateful for experimental proof, but in his view Hughes' experiments had furnished nothing because they were so ineptly conducted.

In a further note to *The Electrician* he lectured Hughes on how to do experiments – 'to be of any use we must know what we are measuring and verifying' – and went on to enlighten him and readers about the skin effect before finishing with a sarcastic thrust.[5]

> Having been, so far as I know, the first to correctly describe the way the current rises in a wire, viz, by diffusion from its boundary, and the consequent approximation, under certain circumstances, to mere surface conduction; and believing Prof. Hughes's researches to furnish experimental verifications of my views, it will be readily understood that I am specially interested in this effect and I can (in anticipation) return thanks to Prof. Hughes for accurate measures of the same, expressed in intelligible form …

Needless to say, Hughes did not take up the invitation.

The Hughes episode had raised Heaviside's hackles but the second incident, in the spring of 1887, drove him to a fury that simmered for the rest of his life. It concerned one of his great gifts to the world, the recipe for a distortion-free telephone line, and the villain of the piece was Hughes' friend and Oliver's old adversary, William Preece. As we have seen, Heaviside had already come close to discovering the formula for ridding telephone systems of the distortion that blighted their progress, but at that time no one, not even Oliver himself, knew that such a thing was possible. As often happens, the opportunity for discovery came by chance.

Oliver's brother Arthur had proposed a radical redesign of the Post Office's Newcastle telephone network. His idea was to adopt what was called a bridge system, putting the telephone receivers in parallel rather than in series. It was a big project and he asked Oliver to work out the underlying theory. In fact, this was familiar ground. In Arthur's scheme each telephone would be, in effect, a deliberately inserted leak, or fault, between the outward and return conductors and Oliver had already written at length about the effect of leaks. Early telegraphers had found to their surprise that leaks sometimes *improved* the quality of received signals and Oliver had shown that there were good theoretical reasons why this should be so. He had even suggested building deliberate leaks into lines – this would weaken the received signal but, more importantly, it would also reduce distortion. What held for the telegraph should apply even more strongly to the telephone, where distortion was even more troublesome, so Oliver enthusiastically supported Arthur's scheme. Each telephone set would be doing double duty, serving the customer and improving the quality of sound for telephones further down the line. Of course it couldn't go on for ever: with more and more telephones in parallel, like rungs on a ladder, the signal would eventually become too weak – but the layout was a great improvement on the existing one, in which the telephones were strung in series, like wagons in a train. Arthur's scheme was derided at first by the old guard but its performance soon silenced the critics and it became the standard method, not only in Newcastle but everywhere. A big advance, but a greater one was to follow.

It started with a trivial-seeming observation. The effect of a leak was opposite to that of a resistance in the line. A resistance acted like a partial dam, increasing the line's tendency to hold its electric charge and restricting the current flow, but a leak provided an extra escape channel for the charge and allowed more current to flow from the battery. Could things be arranged so that the two effects cancelled one another? To investigate the possibility, Oliver conducted a thought experiment.

He pictured an ideal transmission line – one with inductance and capacitance but no resistance or leakage. A signal would simply pass as a wave along such a line, with no distortion or weakening. He then introduced both a small resistance and a slight leak at a single point in the line and considered what effect they would have on a passing wave. Each would reflect part of the wave back up the line but the two reflected components would tend to cancel one another. By its partial dam effect the resistance would reflect the current negatively, thus reducing the current and building up the electric charge on the line. The leak, on the other hand, would reflect the current positively, increasing the current upstream of the leak and reducing the charge on the line. By choosing suitable values for the resistance and the leak, the two reflected components could indeed be made to cancel one another. In passing the resistance and the leak, the wave would then be reduced slightly in strength – part of its energy being converted into heat – but it would keep its shape. There would be *no distortion*.

Ideal lines did not, of course, exist. Real lines had resistance, because no metal was a perfect conductor, and they also had leakage, because the insulation could never be made complete. And both the resistance and the leakage were distributed continuously along the line. To approximate the real-life situation, Oliver thought through what would happen if a square pulse encountered several of his resistance/leak elements, one after the other, as it travelled down his otherwise ideal line. At this stage he didn't use mathematics but sketched pictures of the pulse as it went through. The process was intricate: at every resistance/leak element the pulse would split, part going through and part being reflected; then both parts would split again when they came to another element, and so on. The effect of resistance alone on the square pulse would be to round off its sharp corners and to give it a long tail, like a bridal train. But if the resistance and leakage effects were brought into balance there would be no rounding and no tail: the pulse would become weakened as it travelled but it would keep its original shape. And this would still happen if the number of resistance/leak elements were very large, or even if the whole line were made up of such elements. In other words, if the line were a real one with continuously distributed leakage and resistance it would transmit signals without distortion if the resistance and leakage effects could somehow be brought into balance.[6]

He scented discovery. Nothing could stop signals becoming weaker as they passed along the line but now he knew that he should be able to find a formula for a condition that would allow them to travel without distortion. It would involve the four characteristics of the line: its resistance R, leakance G, inductance L and capacitance C – all expressed per unit length. He turned to his differential equation for a transmission line and, now that he knew what he was looking for, the formula almost jumped off the page. The way to banish distortion was to simply make the ratio of resistance to leakance equal to that of inductance to capacitance. In symbols:[7]

$$R/G = L/C$$

or, rearranging things slightly:

$$L/R = C/G$$

This was the simple recipe for making a telephone line that would send recognisable speech over long distances. It also showed very clearly why distortion was such a problem in existing lines: they had values of L/R much smaller than C/G. To cut distortion down to a tolerable level the two ratios would have to be brought closer together.

So L/R had to be increased, or C/G reduced, or both. Engineers already regarded resistance as an evil to be minimised and the more enlightened of them knew enough about capacitance to know that it, too, should be kept as small as possible. The resistance R could be reduced further, at considerable cost, by using heavier wires or by making them from copper rather than iron, but L/R would still be much smaller than C/G. And there was little scope for reducing the capacitance C. What, then, about increasing the leakance G? Oliver had shown that leaks, within reason, could be beneficial but they weakened the signal and the effect of increased leakance would be cumulative along the line. If you increased the leakance enough to get close to the distortionless condition, the signal would become far too weak to be useful by the time it got to the end of a long line. Only one possibility remained – to increase the inductance L. But this was no meagre Hobson's choice; it was plainly the right course. There were no significant drawbacks, only benefits. Time, surely, to tell the world about it.

Buoyed up by their successes, Arthur and Oliver wrote a substantial paper on the bridge system of telephony and its implications. Oliver's ideas on how to make long-distance telephony possible were developed at length in a technical appendix. Before sending the paper to a journal, Arthur had, by Post Office rules, to clear it with the Assistant Engineer-in-Chief. A formality, one would have thought, especially since it was an enterprising and innovative paper that would have reflected great credit on the Post Office. But the Assistant Chief was incensed; he turned the paper down and even refused to return it to the authors. His name was William Preece.

From our distant viewpoint, Preece's action seems one of plain stupidity and spite. Yet he was not a bad man, nor, in most things, a foolish one. A brilliant organiser and manager, he held the respect of his colleagues and was revered by many of the junior staff. He carried his authority with ease and style, and had become a highly effective public speaker, combining assiduous preparation with a flair for improvisation when needed. When he became President of the Institution of Electrical Engineers a few years later he was described by his predecessor as 'undoubtedly the most popular man in the profession'.

Born in Caernarvon, North Wales in 1834, William Preece was the son of Richard, a remarkable man who began working life as a schoolteacher and Methodist preacher and became Mayor of Caernarvon, by which time he had changed his profession to banking and had become proprietor of the town's newspaper. Richard's great achievement was to clear Caernarvon of huge debts built up by its former Corporation and to make the town an attractive place to live. Unfortunately, he succumbed to the temptations of financial speculation. At first things went well. When William was eleven the family moved to a fashionable house in London and he was sent to King's College School. When the school opened a Military Department, William enrolled with a view to getting a commission in the army but his father Richard's financial speculations began to go awry and that scheme had to be abandoned: in those days commissions had to be purchased and young officers needed a private income to supplement their modest pay. Eighteen-year-old William left the school, ostensibly to study at home and look for a job, but his diary at this time paints a picture of a young man-about-town, enjoying parties and the theatre, playing billiards and developing a taste for cigars.

Reality struck after a year or so when he took a job with the Electric Telegraph Company. The chance came through family friends Edwin and Latimer Clark, both distinguished engineers with the company. At his interview he airily asked for £150 per year – 'Made a fool of myself', as he put it later. They took him on for half that sum and he joined four other young clerks in a small office at the company's headquarters in The Strand. His friendly exuberance, which so far seemed to have opened all doors for him, failed to impress his new workmates. His diary records 'Bad commencement – obliged to find out everything for myself.' Find out he did, and from then on he was determined to make a success of his career.

He soon became an assistant engineer and one of his early assignments was to help Michael Faraday and the Astronomer Royal, Sir George Airy, with some experiments on signal retardation in submarine cables. Airy had wanted to send precise Greenwich Mean Time signals by telegraph to Paris and was disappointed to find that the transmission was too sluggish for the purpose. Preece began to make a reputation as a resourceful and innovative engineer and within three years was Superintendent of the Company's Southern District. As a sideline he designed and built a signalling system for the London and South Western Railway that set a new standard for safety on trains.

When the Post Office took over all mainland telegraph operations in 1870 he was a strong candidate for promotion and was appointed Engineer in charge of the Southern Division. He was by then an experienced author of technical papers and in 1872 published one on duplex telegraphy, only to see his views ridiculed the following year in the *Philosophical Magazine* by a 23-year-old telegraph operator called Heaviside. Preece was, by now, accustomed to deference from young men and assault came as a shock. As we have seen, he wrote to Engineer in Chief R.S. C about this 'most pretentious and impudent paper' and Culley replied that they w 'try to pot Oliver somehow'. One way of 'potting' him would be to recruit hi that he could be kept under control, and this thought may have prompted Pree offer to take Oliver onto the Post Office's payroll when left the Great Northern 1874. Oliver spurned the invitation, of course. The scene was played out again 1881. Using Arthur as intermediary, Preece offered Oliver a job at the good sala of £250 per year. It would have been a different kind of 'potting' – the job was 400 miles away with the Western Union Company in America – but Oliver didn't wan anything to do with this one either.

William Preece's serene upward progress through the Post Office ranks had suf-fered a slight check a few years earlier. He had set his sights on the top engineering job and an opening came when Culley retired from ill health in 1878. But the post went instead to his long-term and slightly younger rival, Edward Graves. Unluckily for Preece it happened to be a time of forced economy in the Telegraph Department, and Graves was famous for his frugality. By what seems in hindsight a heavy irony, it was Preece's reputation as an enterprising innovator that may have cost him the job. He did get promotion in the shake-up after Culley's retirement but it was only to Assistant Engineer-in-Chief. As a natural leader, he must have been doubly disappointed to find himself in what was mainly a staff role, outside the principal line of command: his responsibilities included such things as engineering methods, technical efficiency, and research and development. To his credit, he bore no grudge, gave Graves loyal support, and tackled the new duties with all his customary vigour. His reputation flourished: he became President of the Society of Telegraph Engineers in 1880 and was elected a Fellow of the Royal Society the following year.

The telephone was making its first appearance in Britain and Preece persuaded the more cautious Graves to strike up an arrangement with Alexander Graham Bell for use of his patent. No sooner had the Post Office dipped its toe tentatively in the water than the tide rushed in. British subsidiaries of the American Bell and Edison Companies each set up a telephone exchange in London and began a private war, fighting over rooftop routes for wires and sabotaging each other's lines. At the same time they were squabbling in the courts over patent rights. The mutual damage from all this was so severe that in 1880 they decided to shake hands and pool their British operations by forming the United Telephone Company. They had another battle on their hands though: the Post Office accused them of infringing its sole right to operate *telegraphs* in Britain. The case went to a senior judge, who had to decide whether or not the telephone was, legally speaking, a form of telegraph. In a hearing that could have come from a Gilbert and Sullivan comic opera, Preece gave evidence that indeed it was a form of telegraph, but found his own words from various publications quoted by the defence to show that it wasn't. After eight days the Post Office won its case and all private telephone companies had to apply for licences.

The early years of the telephone in Britain were anarchic and acrimonious. Com-
complained about the terms of their licences and blamed their frequent failures
.. Preece would have liked to buy them all out and estab-
couldn't persuade senior colleagues and ministers to take
end he had his way, but for the present he pressed merely
trunk lines between cities, which carried the relatively long-
were the lines which suffered most from signal distortion and
he knew how to minimise it.

ution was to replace the iron wires in both telegraph and telephone
pper ones. They had lower electrical resistance than iron wires
more expensive, and to persuade Graves and the Treasury that the
well spent he needed a very strong case. Where copper wires had
did indeed improve signal transmission but the evidence was patchy
quantify. To strengthen his argument Preece sought a compelling new
reason for the superiority of copper wires, and found one. Unfortunately,
g. Over the years he had come to believe that self-induction was the *bête*
telegraph and telephone circuits, to be hunted down and, as far as possible,
d, because it slowed everything down by 'dragging on the current'. Copper
were better than iron ones, he maintained, because they had lower inductance;

in fact copper had, for practical purposes, no inductance at all! He had proved this by experiment and if it contradicted Maxwell's theory then Maxwell must be wrong. He wrote:[8]

> Followers of Maxwell have placed too great reliance on a formula of his which gives a coefficient of self-inductance greatly in excess of what is met with in practice, and it is questionable whether Maxwell's formula is based on a wrong assumption.

Such was Preece's confidence that he put forward his favourite rule of thumb as *the* law governing the maximum distance for transmission of useful signals. It was his own, misapplied, version of William Thomson's 33-year-old '*KR law*' and he wrote it as:[9]

$$x^2 = A/KR$$

where x was the maximum distance, and K and R were respectively the line's capacitance and resistance, both expressed per unit length*. A was what can fairly be described as a fiddle factor: it varied according to the type of line, and its setting required expert judgement – his own in fact. The formula may have looked impressive but its theoretical credentials were weak. It was an incorrect application of Thomson's theory, which was valid for slowly worked submarine cables but not for Preece's high-speed landlines. Even on Preece's own terms the formula was deficient. It took no explicit account of inductance, the very factor he considered the greatest hindrance to long-distance transmission – inductance came into his calculations only as a contributor to the mysterious catch-all factor A. Preece gave values of A for copper and iron wires but didn't explain how he arrived at the figures.

His campaign for copper wires became a topic for debate among members of the Society of Telegraph Engineers and Electricians. Among those who opposed it was Professor Sylvanus P. Thompson, a young but highly respected and influential physicist and engineer. Thompson said there was little point in trying to improve the performance of transmission lines: what was needed was better sending and receiving gear. To Preece this seemed like sedition and he rashly attacked Thompson on theoretical grounds. In fact Preece was right – the greatest potential improvement did lie in better lines and changing from iron to copper would have been a good start – but by venturing into theory he had exposed his weak flank. Thompson struck back.[10]

> I should recommend Mr. Preece for the future to avoid mathematical arguments. My only reason for replying to his extraordinary statements about mathematical theory is that I do not let them pass unchallenged for fear that other members of the Society would otherwise think that Mr. Preece's ideas on this topic were either authoritative or intelligible.

Preece's standing was so secure that Thompson's battering ram caused barely a tremor. But the criticism was a personal affront, an assault on Preece's dignity, and it hurt. Resentment burned. He saw no way of getting back at Thompson but was ready to strike out at any other attacker who came within range. Exactly at this time, the Heavisides' paper landed on his desk. Like an ineradicable pest, Oliver had returned to torment him – this time by asserting that self-induction, far from being a bane to

---

* The symbol K was then commonly used for capacitance.

telephony, was actually beneficial. The idea was preposterous: it was like saying that the way to make a horse run faster was to tie its legs together. And, by implicitly repudiating his own widely publicised views, the paper was an attack on his personal authority and reputation. He could not possibly allow it to be published, especially as it would undermine his case for copper wires, which he had based largely on the claims that inductance was a burden and that copper wires were free of it.

Ironically, Oliver had facts at his fingertips that could have *helped* Preece's cause. One reason iron wires were inferior to copper ones was the skin effect. A current always began on the outside of the conductor and at the high frequencies in telephone signals the high magnetic permeability of iron prevented the current in iron wires from penetrating much below the surface. This greatly reduced the effective cross-sectional area of the conductor and increased its resistance accordingly. The skin effect in copper wires was far less marked and so their well-known advantage over iron wires – their lower resistance – was even greater at high frequencies.

Preece not only rejected the paper but refused to return it to the authors. He was not, on the whole, a vindictive man and such an act of spite was uncharacteristic, but he and Oliver always seemed to bring out the worst in one another. Arthur, as usual, was caught in the middle. Not wishing to imperil his career, he submitted to a castigation from Preece, apparently without protest, and even deferred to Preece's views on the evils of self-induction. As Oliver remarked, 'He understands his subordinate position.' There was no cooling of affection between the brothers – Arthur remained the protective elder brother and Oliver felt no sense of betrayal – but they never again wrote a joint paper.

Oliver's anger burned long and deep – even after Preece's death many years later he continued to pour scorn on his memory. For now, the imperative was to tell the world about the dastardly act of censorship. Frontal attack was the only tactic Oliver knew and he sent a letter to *The Electrician* that must have put the wind up its harassed editor. Biggs replied:

> I would use your letter if I could, but it is dangerous in the present state of the law, for however true it may be and however necessary it is libellous. ... Its intention is to show up P. and no matter how justifiable, it is a libel. Candidly, then, I am afraid to use it.

This was not good enough for Oliver, who pressed again for his letter to be published, but Biggs was immovable.

> I am afraid you don't quite understand the peculiar state of the law of libel, nor the ... worries which the possibilities of an action give to all concerned ... It is not the author alone then, but a number of people who suffer, no matter how true the facts of the case or how necessary the exposure.

But nothing could shake Oliver off. He continued to harry Biggs throughout the summer of 1887 and, for good measure, tried another journal, the *Electrical Review*. They, too, feared libel action and declined to publish his exposé of Preece. Desperate to find someone to take up his cause, he wrote to Sylvanus Thompson, the professor who had ridiculed Preece's attempts at mathematics. At first Thompson could not believe what he read, but after Oliver had assured him it was true he wrote back:

> It is monstrous that an official of a Society should misuse his position in a Society to stifle the publication of scientific truths against which he has happened ignorantly, arrogantly, and openly to pronounce an adverse opinion, lest the publication should expose the ignorance and shallowness of his views.

This was music to Oliver's ears, but whatever protest Thompson made within the Society of Telegraph Engineers and Electricians it was not enough – a few years later, after the Society had become the Institution of Electrical Engineers, they again elected Preece as their President.

What about telling the world of the formula for getting rid of distortion from telephone lines? Oliver managed to work it into his regular contributions to *The Electrician* but it was not his style to make a grand announcement. It appeared in June 1887, with no special billing, towards the end of Section 40 of a series of papers with the title 'Electromagnetic Induction and its Propagation'. Oliver's blood was up and in the first draft he had savagely attacked Preece. Biggs nobly upheld Oliver's right to criticise Preece's scientific views but made him tone down the personal abuse. This is what appeared, still dripping with sarcasm.

> Sir W. Thomson's theory of the submarine cable is a splendid thing. His paper on the subject marks a distinct step in the development of electrical theory. Mr Preece is to be congratulated upon having assisted at the experiments upon which Sir W. Thomson based his theory; he should therefore have an unusually complete knowledge of it. But the theory of the eminent scientist does not resemble very closely that of the eminent scienticulist.

At last he had struck home. Reinforced by Oliver's frequent repetition, the epithet 'scienticulist' began to cling to Preece, and his reputation, at least in scientific circles, began to wane. But for the present he continued to make pronouncements that carried authority even when they were absurd. Some of Preece's followers drew the conclusion from his *KR* law that the speed of an electric current in a telegraph or telephone line was inversely proportional to the square of the length of the line. Oliver licked his lips and wrote:[11]

> Is it possible to conceive that the current, when it first sets out to go, say, to Edinburgh, *knows* where it is going, how long a journey it has to make, and where it has to stop, so that it can adjust its (scienticulist) speed accordingly? Of course not; it is infinitely more probable that the current has no choice at all in the matter, that it goes just as fast as the laws of Nature, pre-ordained from time immemorial, will let it; and if the circuit be so constructed that the conditions prevailing are constant, there is every reason to expect that the speed will be constant, whether the line be long or short.

Oliver was in full flow. In parallel with his *Electrician* articles, he had been writing a series, 'On the Self-Induction of Wires', in the *Philosophical Magazine* but after submitting Part 8 in July he was devastated to have it rejected. The series was originally to have had four parts and the magazine had, in truth, been remarkably helpful in extending it to seven, but Oliver didn't see it that way. As if this was not enough, there was ominous news from *The Electrician*: Oliver's remarkably supportive editor, C.H.W. Biggs, had resigned. There is little doubt that Biggs was, in effect, sacked, and that his championing of Oliver had much to do with it. He later revealed that he had continued to publish Heaviside's articles in the face of strong opposition from the editorial board and even from his own staff.

A month after Biggs' resignation, the new editor, W.H. Snell, wrote to Oliver.

> ... although I rate the value of your papers very highly indeed, I much regret I have to tell you that I have been unable to discover that they are appreciated by anything like a sufficient number of our readers to justify me in requesting you to continue them. I have taken special pains to inform myself upon this point and after inquiring in the quarters where students might confidently have been expected I have not been able to discover any ... Pray accept my expressions of sincere regret.

Was Preece behind this? Oliver thought so. And what 'special pains' had Snell taken? We don't know. One story is that he gave instructions for a loose sheet to be put in each copy asking Heaviside readers to reply but the printer forgot to include the inserts. What we can be sure of is that very few people read the papers in detail but a larger number, including Snell himself, could recognise their importance.

Censored by Preece, and with yet another door shut in his face, Oliver could be forgiven for thinking the world was against him. But he was resilient. Reflecting later on the black year of 1887 he said that if his work had continued to be blocked 'then I should have been obliged to take some very decisive measures, and I am a determined character in my way'. Whatever he had in mind, he didn't stop writing. Indeed he confidently began a new series of six papers with the title 'On Electromagnetic Waves'. The experiences of the last few months had awakened a fierce desire for recognition – without it he would continue to be trampled on – and for the first time he wrote with a specific aim in mind. Rather than simply following his muse, he set out in the new series to gain a scientific reputation – to establish himself as a leading theorist of electromagnetism.

Faced with the problem of getting the new series published, he took the boldest possible course and went, as he put it, 'to headquarters'. When, as a young telegrapher, he had been refused membership of the newly formed Society of Telegraph Engineers he had gone straight to the country's foremost scientist, William Thomson, and asked the great man to propose him. Audacity had served him well then and perhaps it would again. Thomson had had a long association with the *Philosophical Magazine* and was now an advisory editor. One favourable word from him would be enough, so Oliver sent him the manuscript of the first two parts of 'On Electromagnetic Waves'.

The strategy was risky. For all his brilliance, Thomson had never understood Maxwell's theory and remained sceptical about it. When Thomson wrote back with some queries Oliver had to decide whether to tell him he was wrong. In for a penny, in for a pound: Oliver told Thomson as politely as he could that he was 'on the wrong track' and that his opposition to Maxwell could prove 'disastrous for progress'.

Boldness won the day. In December the *Philosophical Magazine* accepted one of Oliver's articles and agreed to follow it with his series 'On Electromagnetic Waves'.[12] Then, in March 1888, came the first public recognition Heaviside had ever received in his life. The first paper in his new series had caught the eye of Oliver Lodge, who then looked up Heaviside's earlier work and liked what he saw. In the second of his well-publicised lectures on lightning to the Society of Arts, Lodge, as we have seen, praised Heaviside's 'singular insight' and his 'masterly grasp of a most difficult theory'.

A new phase of Oliver's life began. It came with feelings of relief and joy: he was no longer alone. No wonder he told Lodge 'I looked upon your second lecture when I read it as a sort of special providence!' After fourteen years of solitary work he had found a kindred spirit. Through Lodge he found Fitzgerald. Then came Hertz. And Hertz's brilliant experiments brought about a great swell of interest in everything to do with electromagnetism. Oliver's toil had not been in vain: his time had come.

Chapter 8
Self-induction's in the air
Bath and London 1888–89

The British Association for the Advancement of Science held its 1888 meeting in Bath. The annual gatherings were held in a different city each year and were grand affairs; about two thousand visitors travelled to Bath in September to hear about the latest discoveries and debate the issues of the day. Some had come to hear reports from African explorers and some to listen to a young man called George Bernard Shaw talking about *The Transition to Social Democracy*. But the subject of compelling scientific interest this year was electricity. People thronged to see and listen to Edison's improved wax phonograph, and electrical topics dominated discussion in both Section A (Mathematics and Physics) and Section G (Engineering).

The great and good were well represented. Among the physicists, Sir William Thomson was joined by Lord Rayleigh, who had succeeded Maxwell at the Cavendish Laboratory in Cambridge, and William Preece took his place as President of the Engineering Section. Oliver Heaviside never attended British Association meetings, or any meetings for that matter, but at last his name was becoming known and at Bath he took on the role that became his trade mark – that of the *éminence grise*. But the star of the show this year was another absentee, Heinrich Hertz.

Word had begun to get around of Hertz's discovery but not many people in England read the *Annalen der Physik und Chemie* so Fitzgerald and Lodge were keen to tell the assembled company of the wonderful experiments that had settled decades of speculation and controversy about electromagnetism. The President of Section A was Arthur Schuster, a former student of Maxwell's, and delegates had expected him to start things off with an account of his work on electrical conduction through gases. But Schuster fell ill and Fitzgerald was chosen as acting President. The stage was set and Fitzgerald seized the historic moment. He described how Hertz had, by the 'beautiful device' of his spark-loops, generated and detected waves not only along wires but in free space, and had shown that they travelled at the speed of light, as Maxwell had predicted. This could only happen if the space through which the waves passed acted as a repository and transmitter of energy; the experiments had demolished all theories of action at a distance.

Possibly no scientist ever received more heartfelt acclaim at a conference than Hertz did at Bath. It didn't seem to matter that he was both absent and foreign. As Lodge put it, he had 'cut into the ripe corn of scientific opinion in these islands'. The triumph of the Maxwellians was almost complete – most of the doubters were won over to Maxwell's theory of electromagnetism – but William Thomson remained obdurate. He still took a strictly mechanical view of the universe and thought that Maxwell's idea of displacement currents – fleeting currents in insulators or empty space – was crazy. Much as his fellow scientists respected and loved him, they began to think that he was falling behind the times. But such was his reputation in the country that the national press tended to hang on to his every word. When, at Bath, he remarked, tongue in cheek, that he had softened his critical opinion of Maxwell's displacement currents from 'wholly untenable' to 'not wholly tenable' *The Times* solemnly reported the event as 'of great importance to all interested in electromagnetic theory'.

Many at Bath did become interested in finding out more about electromagnetic theory, especially the latest extensions to Maxwell's work. Word passed around that Oliver's work was way ahead of anyone else's and Fitzgerald made a point of telling anyone in Section A seeking enlightenment to look at 'the writings of Mr Heaviside'.

The occasion that really stirred passions at Bath was a joint session of Sections A and G. Its purpose was to debate the controversial views on lightning conductors that Lodge had expressed in his spring lectures to the Society of Arts but it turned into a highly publicised war of words between 'practice' and 'theory'. Preece had already set the mood in Section G's own meeting by asserting in his Presidential address that the 'practical man', who had 'his eye and mind trained by the stern realities of daily experience, on a scale vast compared with the little world of the laboratory', would not be dictated to by any theorist. Sensing the tension, many delegates from other sections deserted their own discussions to come to watch Sections A and G fight it out.

Preece spoke first. His views on lightning conductors were, on the whole, correct, and Lodge's wrong. Had he presented the case in a straightforward way he could have gained much credit for himself and the practical men he claimed to represent. Instead, he walked into trouble. Apparently heedless of who was in the audience, he launched an attack on people whose obsession with 'mere mathematical development' was a wasteful distraction from the real work of the day. Mathematicians should know their place. For his own part, he had 'made mathematics his slave and did not allow mathematics to make him its slave'. This was too much for Lord Rayleigh, who rose to complain that while listening to Preece 'one felt that "mathematician" was becoming a term of abuse'. Thomson came in behind Rayleigh and this seemed to bring Preece up with a start. In an instant he changed from bulldog to poodle. He wouldn't presume to question the views of 'masters in mathematic'; in fact the whole assembled company were 'students at the feet of their Gamaliel*, Lord Rayleigh'. In an effort to restore his dignity, Preece explained that what concerned him were 'some of those young fellows coming out every year with a smattering of mathematics; they wrote Papers for the technical journals and they thrust upon the electrical world conditions and conclusions arrived at by their mathematics with a coolness and effrontery that was simply appalling'. Oliver's ears must have been burning.

---

* Gamaliel was the apostle Paul's teacher.

On what was intended to be the topic of the session, Preece started well, claiming – correctly, as it turned out – that Lodge was mistaken in assuming that all lightning flashes carried extremely high-frequency oscillations. By the same token, he was right when he denied Lodge's assertion that self-induction played a defining role in lightning protection by confining currents to the skin of the conductor. But he went on to say that Lodge and his friends had 'a mania for self-induction'; that 'lately, self-induction had been brought in to account for every known phenomenon'; and that self-induction was a 'bug-a-boo' to be shunned by practical men. As if this were not enough, he made a risky reference to John Henry Poynting's theoretical work on energy flow, of which he knew next to nothing, and attempted a joke by comparing Lodge to the prophet Balaam who was hired by Balak, king of the Moabites, to put a curse on the invading Israelites but failed to carry out his commission, all the while riding an ass. (The British Association was colloquially known as the British Ass.)

Knowing he was up against a speaker as practised and polished as himself, he should have stayed on safe ground. Lodge had little difficulty showing up his opponent's ignorance of theory and neatly turned Preece's Biblical story round by claiming that his own predictions, like Balaam's, would come true, and that if Balak was the Society of Arts, who had commissioned his talks on lightning, then the ass must be Preece.[1] Knockabout, schoolboy stuff, but he drew the biggest laugh of the day and in its account of the debate *The Times* reported general agreement that 'Professor Lodge had the better of it'.

Preece did himself no favours by following up a few weeks later with a peevish letter to Lodge, saying that in 35 years of work on electricity he couldn't recall a single instance when he had derived any benefit from pure theory. He added that the Atlantic cable project had owed nothing to William Thomson's theory: that all the real work had been done by practical men. Lodge, of course, showed the letter to all his scientific colleagues, including Thomson, and Preece fell further in their estimation.

Back in Camden Town, Heaviside was cock-a-hoop. Maxwell's theory had leapt to prominence and Preece had been taken down a peg. Better still, Preece's tirade on the evils of self-induction had, contrarily, stimulated interest in his own work on the role of self-induction in the propagation of waves. And a new acting editor at *The Electrician*, W.G. Bond, had accepted two of his articles. Brief and jaunty, they both took up themes from the Bath meeting – lightning discharges and practice versus theory. In the latter he took several merry swipes at Preece, for example:[2]

Is self-induction played out? I think not. What is played out is … the British Engineer's self-induction, which stands still and won't go. But the other self-induction, in spite of strenuous attempts to stop it, goes on moving; nay more, it is accumulating momentum rapidly and will, I imagine, never be stopped again. It is, as Sir W. Thomson is reported to have remarked, 'in the air'. Then there are the electromagnetic waves. Not so long ago they were nowhere; now they are everywhere, even in the Post Office. Mr. Preece has been advising Prof. Lodge to read Prof. Poynting's paper on the transfer of energy. This is progress indeed! Now these waves are also in the air and it is the 'great bug' self-induction that keeps them going.

By way of private celebration, he broke into verse. A notebook entry reads:

Self-induction's 'in the air'
Everywhere, everywhere;
Waves are running to and fro
Here they are, there they go.
Try to stop them if you can,
You British Engineering man!
Conceive him (if you can)
The engineering man,
Docking and blocking and burking a Paper
Up in St. Martin's-le-Grand!

Burking was contemporary slang for murdering by suffocation, after William Burke who was hanged in 1829, and St. Martin's-le-Grand was the site of the Post Office's London headquarters.

The world had opened up for Oliver. No longer was he an obscure self-educated former telegraph operator writing articles nobody read; his work was discussed and praised by eminent professors who welcomed him as a colleague. To top it all, he soon received what was tantamount to an official seal of approval at the highest level – a public tribute from William Thomson. And it came in the most potent form possible – in Thomson's Presidential Address to the newly formed Institution of Electrical Engineers, which had replaced the Society of Telegraph Engineers and Electricians.[3] Talking of the theory of wave propagation along wires, Thomson said:[4]

> It has been worked out in a very complete manner by Mr. Oliver Heaviside: and Mr. Heaviside has pointed out and accentuated this result of his mathematical theory – that electromagnetic induction is a positive benefit: it carries the current. It is the same kind of benefit that mass is to a body shoved along a viscous resistance ... Heaviside's way of looking at the submarine cable problem is just one instance of how the highest mathematical power of working and of judging as to physical applications helps on the doctrine, and directs it into a practical channel.

What a man! Thomson was always prepared to use his influence to bring talented people along and had done so three times for Heaviside – first to get him into the Society of Telegraph Engineers after he had been rebuffed as a mere clerk, then to persuade the *Philosophical Magazine* to resume publishing his articles, and finally to tell the professional electrical engineers that they needed to take note of Heaviside's work if they wanted to make progress. It had not mattered that he and Oliver held opposing views on Maxwell's theory, nor that Oliver had given offence with his irreverent comments on archbishops; Thomson recognised the brilliance and integrity of Heaviside's work and felt a duty to commend it. Besides, he probably had an abiding admiration for the sheer spirit of the young man who had walked up to him 15 years before and asked to be proposed for membership of the Society.

Elated, Oliver wrote to Thomson:[5]

> I have just read your Address to the new Inst. of E.E., and write to thank you for your most hearty appreciation of my work in connection with waves along wires. I am ashamed to see it occupy so much space in your Address, and think it must have crowded out your own matter which would have been of greater and more permanent scientific interest. Your

appreciation is most welcome after the continued indifference and (sometimes) opposition I have met with.

As Paul Nahin has aptly observed, this is about as humble as Oliver got in his life.

Thomson had made someone else happy too. C.H.W. Biggs saw Thomson's Presidential Address as a vindication of his own support of Heaviside while editor of *The Electrician*. Standing by Oliver then had cost him his job but he was now editor of *The Electrical Engineer (London)* and this was the perfect opportunity to get things off his chest. In an editorial, he went some way towards explaining why his old journal had stopped publishing Oliver's papers. He was itching to implicate Preece but had to tiptoe round the law of libel. In doing so, and in disguising his meaning with irony, Biggs produced what was probably the least intelligible sentence he ever wrote. Perhaps this didn't matter – many people would be able to read between the lines. With the enigmatic sting in its tail, the editorial ran:[6]

> This may not be the best place to record certain facts, in connection with Mr Heaviside's writings, but we venture to do so for many reasons, not the least of which is the pleasure we feel in having our views corroborated by so eminent an authority as Sir W. Thomson. Putting aside the papers that appeared in the *Phil. Mag.*, articles that were given in *The Electrician* were given in spite of the most strenuous opposition by the proprietors and every member of the staff, except the late editor. Hence possibly the singularity of the sudden cessation of the articles which has so frequently been remarked upon that the reason may not be uninteresting. Previous to Sir W. Thomson's acknowledgment of the magnitude and value of Mr. Heaviside's work, Mr. Preece had called attention to certain parts of it as worthy of the closest attention and forestalling later workers in the same direction.

William Preece was taking a pounding but he never lacked fortitude. There were many strands to his life and it didn't matter much if one got a bit frayed. Besides carrying on his work at the Post Office with undiminished zest, he became the country's most frequently consulted expert on electric lighting. And he got his way on copper wires: after a pilot line from London to North Wales proved a big success, the Post Office changed to a 'copper standard' everywhere. He made many working visits to other countries and was by now an international figure; among the honours that had come his way was *Chevalier de la Légion d'Honneur*. He was promoted to Engineer-in-Chief when Graves died suddenly in 1892, and on his retirement from the Post Office seven years later he was awarded a knighthood.

Preece's life was so full that his feud with Oliver Heaviside soon faded from thought. But things looked different from the other direction: Oliver remained brimful of rancour and looked for every opportunity to let it out. When *The Electrician*'s new editor, A.P. Trotter,[7] sent a friendly letter asking him for more articles, Oliver probably rubbed his hands at the prospect of getting in plenty of jibes at Preece. But Trotter was wise to such tactics and put his foot down.[8]

> I object to contentious discussions, as they are apt to degenerate into a tone that is unsuitable to this paper. I reserve the right to cut or modify such matters.

This drew an extraordinary response from Oliver.[9]

> I never have, and do not intend to have anything to do with contentious discussions, as I understand 'contentious', contending for the sake of contending, about nothing worth

contending about. Legitimate scientific reasoning means that if anyone puts forward views which I consider wrong, and am interested in, and the matter is worth correcting, I have the right to do it, on scientific grounds.

Heaviside the non-controversialist! He knew Trotter's meaning perfectly well – cut out *personal* attacks on Preece or anyone else – and was tacitly conceding the point by resorting to semantics. Trotter was doing him a good service and, in his heart of hearts, Oliver probably knew it, though he still hounded Preece whenever he could. In other matters Trotter handled his difficult writer kindly, for example yielding to Oliver's demands for fortnightly publication whenever possible and generous notice of any cancellation.

The heady taste of recognition had given Oliver a thirst for more. In a rare moment of entrepreneurial zeal he decided to have fifty copies of his series 'On Electromagnetic Waves' printed and bound so that he could send them to leading physicists all over the world.[10] Lodge helped him choose the names. How he raised the money is a mystery. Perhaps he had saved enough over the years out of his tiny income from article fees, or perhaps Arthur helped. At all events the venture went well enough for a notebook entry to record it as 'a good stroke of business'. And much more than this, it helped to bring about events that secured Heaviside's writings for future generations.

The little book led to a big one, or rather two. At Trotter's suggestion, Oliver began writing his new series of articles with a view to their coming out later in book form. This raised the question of what to do about the earlier papers. In Oliver's words, 'as the later work grows out of the earlier, it seemed an absurdity to leave the earlier work behind'. A way had to be found to republish all of his papers up to 1890 in book form. With the possibility of sales in America and Europe as well as Britain, the proposition was now a reasonably attractive one to publishers, and after some debate over who owned rights to what – the papers had been spread among five different journals – Macmillan agreed to take on the job. They started assembling the papers into two volumes, to be published under the title *Electrical Papers*. Many of the papers had been written in instalments, like a Dickens novel, and so fitted fairly neatly into book form. *Electrical Papers* came out in 1892 and the new series Oliver had begun for *The Electrician* eventually became his three-volume treatise on *Electromagnetic Theory*. Both publications are now classics, still in print.

Many people were now keen to meet the reclusive Heaviside. Fitzgerald and Lodge each called to see him in Camden Town early in 1889. Fitzgerald was the first to visit and got lost in the surrounding streets before finding St Augustine's Road. Lodge had no such problem, having lived nearby when a student, but was distressed to find his new friend living in such 'dismal lodgings'. Despite the grim setting they found Oliver a genial companion and would have loved to have introduced him to a wider circle. They and others did their utmost to entice him out, but failed. One invitation, from Sylvanus P. Thompson in January 1989, could hardly have been more pleasantly expressed.

> I know that you live a very quiet life. Do you mind for once a little scientific excitement? On 7th March Oliver Lodge will be staying with me – we live very quietly, and my household is a quiet one – would you join us that evening at dinner, and we would have a three-corner chat afterwards on electrical matters.

One wonders how he could turn down invitations like this from people who appreciated his work and wished him well. He had greatly enjoyed meeting Fitzgerald and Lodge but remained unaccountably averse to any sort of gathering. The trouble seemed to be shyness, but not of a common kind – he had been bold enough when a young man to buttonhole the great Sir William Thomson. Deafness and susceptibility to ailments like dyspepsia and the 'hot and cold' disease may partly explain his voluntary reclusion, but in the main it remains, like other aspects of Heaviside's character, an enigma.

What of his great discovery: how to rid telephone lines of distortion? It was there, in one of his *Electrician* papers of 1887 – make L/R equal to C/G – but nobody paid much attention.[11] Even Heaviside himself did not make great play of it. This seems surprising in hindsight but there were several reasons.

Oliver was by now taken up with his new series on electromagnetic theory. This was an ambitious project. He wanted to build up the whole subject from scratch in his own fashion; ways of reducing distortion in telephone speech would appear again in their proper context but he needed to set the groundwork first. And he had in any case found an important new role for the distortionless condition: it was not only a means of improving telephone communication but a 'royal road' to the understanding of Maxwell's theory.

A road was, indeed, needed. Heaviside had himself explained the complex behaviour of transmission lines using no more than linear circuit theory, in which electric currents supposedly flowed along single-dimensional wires. Yet he was now telling the world that the true source of all knowledge in electricity and magnetism was Maxwell's theory of *three-dimensional* electromagnetic forces and fields. How were the two approaches to be reconciled? Oliver had worked out everything to his own satisfaction years before. Linear circuit theory was simply an immensely useful approximation that worked well as long as frequencies were low enough for field phenomena such as the skin effect to be ignored. He had written paper after paper on the topic, explaining that electrical energy travelled in the space *alongside* the wire and that the only flow of energy in the wire itself was *inwards*, at right angles to the direction of the wire. But these ideas were still foreign to the vast majority of people who actually worked with electricity, and thinking about the distortionless transmission line gave Heaviside inspiration for a new and vivid way of getting the message across.

What was a distortionless line but a means of transmitting Maxwellian waves in the direction of the line? They would be, in effect, plane-fronted waves and this meant that the wave equations, and Maxwell's equations themselves, would act in only one dimension – that of the line. What was more, these equations turned out to have the same form as those derived from circuit theory. The correspondence was physical as well as symbolic: the electrical field force corresponded to the line voltage and the magnetic field force to the line current. Engineers who knew some circuit theory could now use that knowledge to gain an understanding of electromagnetic fields. Along an ideal transmission line – one with no resistance or leakage – the waves would travel with no weakening or distortion. Along a line in which resistance and leakage were in balance waves would be weakened as they travelled along but would preserve their shape; they would not be distorted. All this could be shown with only simple mathematics because resistance and leakage tended to reflect the waves in opposite ways, so that when they were in balance the reflections cancelled one another out. And by generalising from the distortionless line, more complex cases could be considered, still without resorting to heavy mathematics. This way, students could get a good intuitive feel for the subject. As Heaviside put it:[12]

> The mathematics was reduced, in the main, to simple algebra, and the manner of transmission of disturbances could be examined in complete detail in an elementary manner. Nor was this all. The distortionless circuit could itself be employed to enable us to understand the inner meaning of the transcendental cases of propagation, when the distortion caused by the resistance of the circuit makes the mathematics more difficult of interpretation.

Using the distortionless line to introduce people to Maxwell's theory was a brilliant insight, but one which distracted attention from its role in practical telephony.

Another reason for not trumpeting the cure for telephone speech distortion more loudly was that with Preece in charge of trunk telephone lines there was no chance of getting the Post Office to take up the idea. Preece would go over Niagara Falls in a barrel rather than agree to increase the inductance in his lines. Oliver and Arthur thought about applying for a patent but it seemed like expense to no purpose. By the same token, there was little point in suggesting specific ways to increase the inductance if no one was prepared to test them. In his 1887 paper Oliver had confined himself to one suggestion: that in submarine cables the insulation between the central wire and the outer sheath could be impregnated with fine iron dust.

He had published no corresponding suggestion for land-lines but did discuss one possibility with William Thomson.[13] It was the simplest possible method: just insert coils of high inductance and low resistance in series at intervals along the line. As long as the intervals were short compared with the wavelength of the transmitted waves, the line would behave almost as though the inductance of the coils was spread evenly along its length and one could get close to the true distortionless condition. As a boy Oliver had found he could make ripples travel better along his mother's washing line by tying knots in the rope. The coils in the telephone line would perform a similar function, adding momentum to help the waves travel smoothly and keep their shape. It was a highly practical idea, but there still seemed to be no point in applying for a patent, at least in Britain. And, perhaps because the method seemed so obvious once the distortionless condition was known, he didn't publish it until 1893. Even then, he didn't put it in the shop window but included it only towards the end of a long series of papers on the 'Theory of Plane Electromagnetic Waves'. As we shall see, there were reasons for this apparently uncharacteristic reticence, but it may have cost him dear.

Thoughts of what might have been were to trouble him later – not so much from lost riches as from lost recognition – when his key part in tackling distortion became obscured in the scramble for patents and royalties. But for now he was on top of the world. He had sent his 'little book' on *Electromagnetic Waves* to Heinrich Hertz and the two men soon became good friends by a lively exchange of letters. Hertz had also set out to simplify Maxwell's theory and, amazingly, had come up with the same four equations as Heaviside.† The coincidence was all the more remarkable because Hertz had taken a completely different approach. He had started from Helmholtz's version of Maxwell's theory, which, in Heaviside's view, was not true Maxwell at all because it clung to action at a distance concepts that Maxwell had spurned. But by a subtle process of successive approximations Hertz eliminated the action-at-a-distance effects, leaving himself with a Maxwellian electromagnetic field. To arrive at the final equations he chose to abandon the electric and magnetic potentials in favour of forces, just as Heaviside had done.

The discovery that his own theoretical work had been matched, or even surpassed, by someone he had never heard of must have come as a shock to Hertz. But he was a man of generous spirit and any envy was swamped by a sense of fellowship. The two men exchanged warm compliments. Hertz told Heaviside he believed:[14]

> ... that you have gone further on than Maxwell and that if he had lived he would have acknowledged the superiority of your methods.

In return, Heaviside thanked Hertz for killing off the action-at-a-distance theories.[15]

> I recognised that these theories were nowhere, in the presence of Maxwell's, and that he was a heaven-born genius. But so long as a strict experimental proof was wanting so long would these speculations continue to flourish. You have given them a death blow.

This letter must have made Hertz wince by reminding him of the damage he had done to his mentor Helmholtz's standing, but its truth was inescapable. A sign of the secure and genuine friendship between Hertz and Heaviside was that each felt free to criticise the other's work. For example, Hertz laid into Heaviside about the need to make his papers easier to follow.[16]

> The fact is that the more things became clearer to myself and the more I then returned to your book, the more I saw that essentially you had already made much earlier the progress I thought to make, and the more the respect for your work was growing in me. But I could not take it immediately from your book, and others told me that they could hardly understand your writing at all, so I felt obliged to give you warning that you are a little obscure for ordinary men.

Hertz had almost become an honorary Englishman and it was no surprise when the Royal Society awarded him its Rumford Medal in 1890. When they invited him to London to receive the medal he accepted, although with characteristic modesty he didn't tell his colleagues in Karlsruhe the purpose of the visit. He had a hero's welcome and met all the leading physicists but one. On a free evening Lodge and Fitzgerald met Hertz for a private dinner at the Langham Hotel and when they sat down they must have felt the presence of a metaphorical empty chair. Heaviside would surely, for once, have left his room and joined them but he was, by this time, living two hundred miles away.

Oliver's brother Charles had moved to the seaside resort of Torquay at the age of nineteen to work as an assistant in a music shop. The business had done well; he was

---

† Hertz would have needed twelve equations – four triples, each with components in the x, y and z directions – but he became an early convert to Heaviside's vector system and so was able to reduce the number to four.

now the senior of two partners and had opened a second shop in nearby Paignton. He had married a local girl, Sarah Way, and they lived with their five children in a flat over the first shop. Above the new shop was another flat. It was spacious and would be a perfect home for his parents Thomas and Rachel, and for Oliver if he wanted to come. Thomas and Rachel were now in their seventies and in poor health – good reason for them to move out of smoky London to the fresh air and sunshine on the Devon coast. It was also a good reason for Oliver to come: he would then be on hand to act as nurse when needed. In fact, Oliver had little choice as he didn't earn enough money to live in tolerable comfort on his own. In any case he had no special liking for London; he could write his papers and letters just as well in Paignton and perhaps the better climate would help to keep the 'hot and cold' disease at bay. In the autumn of 1889 Thomas, Rachel and Oliver moved to Paignton.

Chapter 9
Uncle Olly
Paignton 1889–97

Torbay lies on the part of the South Devon coast that faces east and so is sheltered from the prevailing westerly winds. Its ten miles of shore curve like a horseshoe from Hopes Nose down to Berry Head. The hilly resort of Torquay in the north faces the fishing port of Brixham in the south, and Paignton, with its flat, sandy beaches, sits between the two. During the nineteenth century the bracing but mild climate had attracted new residents, and the local industries of farming and fishing had been supplemented by tourism – the area was well served by the Great Western Railway and was being promoted as the English Riviera. After Camden Town it must have seemed close to heaven.

The move broadened Oliver's social horizon. For one thing, he was cast for the first time in the role of uncle. His brother's five children no doubt regarded him as a bit of an oddity but they took to him and shared some of the same pleasures. The youngest daughter, Beatrice, wrote to a friend many years later:[1]

> He used to be very merry with my brothers and sisters and me. I remember in the big upper stock-room of my Father's music shop, how with my Father playing a march, Uncle Olly, at the head of us, would march around in and out among the pianos (perhaps a dozen or more), we hanging on to his coat tails in a row, one behind the other.

There was music all around. The Salvation Army band used to play on the corner near the flat where Oliver lived and their enthusiastic performances of 'Onward, Christian Soldiers' were not always appreciated when he was trying to work. But, on the whole, music was a joy. Several of the children and their friends were gifted musicians and one of the best was the eldest daughter Rachel's young man, Fred Williams. Fred saw quite a lot of Oliver because he used to call at the Paignton flat when Rachel came over in the evenings to cook supper for her grandparents. He recalled how he was kept busy.

> Oliver was very fond of the best music, especially Beethoven, whose piano sonatas he was never tired of hearing, and I used to struggle through the least difficult ones for his entertainment.

Now that he had Fred and others to play for him, Oliver seems to have abandoned his own attempts to master the piano. Mechanical pianos were soon to start appearing in Charles' shop and, as we shall see, Oliver took to them with gusto. Meanwhile, there was another new invention to brighten his life – the safety bicycle. Unlike earlier machines such as the penny-farthing, the safety bicycle could be ridden by anybody, even ladies! The pedals were in a convenient position and drove the back wheel by a chain while pneumatic tyres cushioned the ride. To stop the machine you squeezed a lever on the handlebars that caused a spoon-shaped piece of metal to be pressed against the front wheel. A craze for bicycling swept the country and Oliver was one of its keenest followers. It took him along the coast and out into the steep lanes of Devon. Sometimes his niece Rachel went with him. On one of their trips Oliver had difficulty pedalling up the hills after discovering the effects of scrumpy – strong local cider. Uphill cycling was always hard work but downhill was a different matter. The early bicycles had no free-wheel device so downhill speed was limited by how fast you could turn the pedals. Unless, of course, you took your feet off. Such dangerous bicycling was known as 'scorching' and Oliver became an expert. He had footrests fitted to the front forks so he could rest his feet there while the pedals whirled round on their own.

Oliver's working routine remained much as it had been for the past fifteen years but with the difference that he was now someone whose scientific views were sought and respected: there were many more letters to write and to answer. In January 1891 he had one from Oliver Lodge asking if he would agree to be put forward as a candidate for fellowship of the Royal Society. Lodge and Fitzgerald had arranged for twelve other Fellows to support his candidature, including Thomson and Poynting. Helped by his new friends, Oliver Heaviside had advanced in only two years from obscurity to the very threshold of the British scientific establishment. Progress indeed, but, as usual, Oliver took a different view of the situation from everybody else: his first thought was to say no! At least he had the sense to consult Arthur, who told him not to be daft. This is the reply that then went to Lodge.[2]

> In my draft reply to yours of the 27th I advanced 7 reasons for declining your offer! But on submission to my brother, although he has not demolished them, yet he has brought some weighty reasons the other way. I will not, therefore, trouble you with them, except one.
>
> Is it not a fact that a Candidate for the R.S. may be down on the list for years? Nothing would be more disagreeable to me. I would not be a Candidate if there were any chance of refusal at the first. *If* a man is good enough, and had shown it, why should he be a supplicant, so to speak?
>
> If, on the other hand, you could assure me that there would be no difficulties in the way, and that immediate election was a moral certainty. Why *then*, I should accept your kind offer with the greatest pleasure; and, in any case, I consider it to be an honour to be proposed by you.

And, in a PS:

> It is not an *unmixed* honour to be F.R.S. now-a-days. Look at the Council List!.

One can only wonder at the patience of his friends. Lodge tried to bring reason to bear and had the following reply.[3]

> You must allow for personal equation. Lots of good men supplicate year after year. Why shouldn't I? Am I better than they? The honour of belonging (after rejections) to such an august and dignified body is transcended by the precious relic of self respect that is left in me by not becoming a supplicant.
>
> Look at it this way too. If a man should talk to me about the ridiculous absurdity of my not taking the R.S. and its rules as I find them, or adapting myself to them, especially

as there is no reason to suppose they will alter them in my favour, I should say, 'The same argument applies to all ancient institutions, ways, manners, customs, etc.; and if followed universally would result in universal stagnation and eternal persistence of the unfit.'

Somebody must decline to adapt himself, and not prefer the old way, like the fishes St. Somebody preached to. And somebody always does. In fact, a good many somebodies do. They are foolish and eccentric, no doubt; but they prepare the way for desirable changes.

A few paragraphs after this philosophical passage came what was probably the real crux of the matter.

My remark 'Look on the Council List' referred to one name in particular; little less than a scandal to be there.

Preece again! Perhaps Oliver feared that his foe may still have had enough influence to block his entry. He certainly thought the Society had demeaned itself by admitting Preece in the first place, let alone by appointing him to its Council.

At length Lodge prevailed, by telling Oliver that the chance of rejection was very slim indeed, and sent the proposal off with tightly crossed fingers. To everyone's relief Oliver was indeed elected on 4 June 1891.[4] But this was not quite the end of the matter. As was customary, the Society's Secretary sent him a copy of the Statutes with a request to come to London for the formal inauguration. He might as well have asked Oliver to swim the English Channel. It was Lodge who again bore the brunt of Oliver's reaction.[5]

The Secretary R.S. has sent me a sort of Habeas Corpus:-

Yet one thing More
Before
Thou perfect Be
Pay us three Poun'
Come up to Town
And then admitted be
But if you Won't
Be Fellow, then Don't.

I don't object to the three poun'; but the exhibition clause is quite new to me. It is made a *sine qua non* of admission to fellowship. Now it is one thing to go to the Society because you have a right to go, *qua* Fellow, but quite another thing to be ordered to attend to be admitted as a Fellow, else null and void. Is it serious, or shall I just let it slide and take no notice?

Like many societies, the Royal Society was fond of its rituals, and it *did* take the admission rule seriously. But nothing could induce Oliver to travel to London. And he had his way. Perhaps Lodge interceded; at all events the Society bent its rules and Oliver was admitted.

Fellows could have papers published in the Society's journal – quaintly called the *Proceedings* – with no need for refereeing, and Oliver lost no time in taking advantage of the privilege. He was proud of his unconventional but highly effective operational calculus, by which difficult differential equations could be solved as if by magic, and decided to write a substantial treatise about it, putting together all that he had done. One aim, naturally, was to spread the word about the spectacular success of the new method in solving real problems but another was to invite mathematicians to share his fascination with the subject and to probe its mysteries. He put it this way in his introduction:[6]

> I have ... convinced myself that the subject is one that deserves to be thoroughly examined and elaborated by mathematicians, so that the method may be brought into general use in mathematical physics, not to supplant ordinary methods, but to supplement them; in short, to be used when it is found useful. As regards the theory of the subject, it is interesting in an unusual degree, and the interest is heightened by the mystery that envelops certain parts of it.

Heaviside's operational calculus did eventually find its way into general use but, as we shall see, it gained entry by the side door after being refused admission at the front. Formal recognition came only *after* ordinary electrical engineers and physicists had found it a godsend – it gave them a way to solve problems that would otherwise have taxed the skill of the best mathematicians. Professor Harold Jeffreys, writing in 1927 about the kinds of mathematical problems that occur in electrical engineering, summed up the case.[7]

> As a matter of practical convenience there can be no doubt that the operational method is by far the best for dealing with the class of problems concerned. It is often said that it will solve no problem that cannot be solved otherwise. Whether this is true would be difficult to say; but it is certain that in a very large class of cases the operational method will give the answer in a page when ordinary methods take five pages, and also that it gives the correct answer when the ordinary methods, through human fallibility, are liable to give a wrong one.

Part I of Oliver's paper 'On Operators in Physical Mathematics' was published in February 1893 and Part II in July. He posted off Part III but the Society sent back a brief and dismissive rejection note. Some Fellows had objected to the liberties Oliver had taken in his mathematics in Parts I and II, and had prevailed on the Secretary to break with custom and have Part III refereed before publication. The chosen referee was William Burnside – an instructor at the Royal Naval College, for whom mathematical rigour went hand-in-hand with military discipline. In such hands the paper didn't stand a chance.

The rejection was a heavy blow. Unlike Preece's act of censorship six years earlier, this one came out of the blue. The Society had welcomed the maverick Oliver into its ranks and humoured his whims but now it had reverted to type, in his view displaying arrogant ignorance. He had hoped that theoretical mathematicians would be so intrigued by the fact that his strange method led to correct solutions that they would set about trying to demystify it by working out a rigorous theoretical base. Instead, they had simply rejected it out of hand.

To understand what all the fuss was about we need to look a little closer into Heaviside's mathematics. As we have seen, his inspiration was to look at Ohm's law in a new way.[8] In the familiar equation

$$V = RI$$

he interpreted the resistance, R, as an *operator* which converts the current, I, into the voltage, V. This deceptively simple step allowed the scope of Ohm's law to be greatly extended. In his generalised version, the law could deal with circuits that had inductance and capacitance as well as resistance and, by the same token, with fluctuating voltages and currents as well as steady ones. One simply replaced the multiplier, R, by a more complex operator which depended on the arrangement of resistance, inductance and capacitance in the circuit. For example, in a circuit consisting of a resistance R in series with an inductance L, one would have:

$$V = RI + L\, dI/dt$$
$$= (R + L\, d/dt)I$$

Or, using Heaviside's symbol p to represent rate of change, d/dt:

$$V = (R + Lp)I$$

In fact, the relationship between voltage and current for any circuit formed of discrete elements of resistance, inductance and capacitance was simply:

$$V = Z(p)I$$

where $Z(p)$, which he called the *resistance operator*, or *impedance*, of the circuit, was the relevant function of the circuit elements and p.[9] This was a differential equation but it looked like an algebraic one and he used the normal rules of algebra to rearrange it into a form from which it could be solved, giving the variation of I over time for a known variation of V, or vice versa. This way, even the most formidable-looking differential equations could be solved with relative ease and Heaviside could work out the response of any circuit to a given input. One device he used was his brilliant *expansion theorem*, by which a difficult function of p could be expressed as the sum of a number of simpler ones. The mathematics here was impeccable but some of his other techniques outraged purists. A few examples will give an idea.

The symbol p represented the mathematical operation of differentiation with respect to time: p operating on distance would give velocity; p^2 operating on distance would give acceleration, and so on. No problem so far, but when solving his equations he took $1/p$ to represent the inverse mathematical operation of integration: in his scheme $1/p$ operating on acceleration would give velocity and $1/p^2$ operating on acceleration would give distance. Heaviside was on thin ice here because the inverse relationship holds only under certain conditions. In his own work he always made sure the conditions were met but he never spelled them out and this made the method risky for adventurous but unwary newcomers.[10]

Sometimes the operational calculus took him into very strange territory indeed. In transmission lines the resistive, inductive and capacitive elements were not discrete but continuously distributed along the line and the operational equations contained the *square root* of p. The operator p meant differentiate with respect to time, so p^2 meant differentiate twice, and so on; but what on earth did $\sqrt{p}$ mean? How could you take the square root of the operation of differentiation? Oliver wasn't put off by such questions; in fact, he was supremely confident on the point.[11]

> The square root of a differentiator occurs in the fundamentals of the physical subject, namely, in the generation of a wave of diffusion. It is necessary and inevitable; also, when studied, it is found to facilitate working.

By way of 'facilitating working' he found, and made much use of, the extraordinary result:[12]

$$\sqrt{p}\mathbf{1} = 1/\sqrt{(\pi t)}$$

Here, t is time and the bold-face $\mathbf{1}$ is his symbol for a function that is zero before $t = 0$ and 1 thereafter, and represents the action of a switch (or telegraph key).* Amazingly, the square root of differentiation is an operation that *can* be performed and the result is a perfectly ordinary function of time. This technique was actually just *within* the boundary of respectable mathematics; it was Heaviside's cavalier use of it that raised hackles. But what really sank Oliver's paper was his use of divergent series.

The Norwegian mathematician Niels Henrik Abel had called divergent series the invention of the devil. But they held a fascination and were being studied warily by some of the best European mathematicians – supping with the devil, using long spoons. Oliver, by contrast, had no fear of the monsters. They cropped up everywhere in the course of his work and he experimented freely, to see what use he could make of them.

What were these divergent series, and why did he feel a need to use them? Like many people with a mathematical job to do Heaviside often found it useful to express a mathematical function as an infinite series, something like:

$$a + bx + cx^2 + dx^3 + \cdots$$

where x was itself often a function of his differential operator p.

Expansion into a series was sometimes the key to solving a problem and it also enabled approximate values of a function to be computed for particular values of x. As long as the series converged, one could get closer and closer to the true value by working out the individual terms, starting from the left, and adding them up. A rapidly converging series could give usefully accurate values from only a few terms. Heaviside spent many hours grinding out figures this way so he could include numerical examples in his papers. He called it 'dreadful work, only suited for very robust intellects', but at least it could be relied upon to give answers.

But not all series converged. We can see this in one of the expansions he liked to use:

$$1/(1 + x) = 1 - x + x^2 - x^3 + \cdots$$

If x is between 0 and 1 the series converges and everything is fine, but if x is greater than 1 the series diverges, and the expansion is no longer valid.† Fortunately, there is an alternative expansion with the opposite characteristic:

$$1/(1 + x) = 1/x - 1/x^2 + 1/x^3 - \cdots$$

---

* Now known as the unit step function, or Heaviside's unit step function, and often written $H(t)$ in his honour.

† For example, if $x = 1/2$ the whole series sums correctly to 2/3, and taking only the first six terms gets you quite close to this, with 21/32. But if $x = 2$ you get the wildly fluctuating series $1 - 2 + 4 - 8 + \ldots$, which bears no relation to the function's value of 1/3.

Unlike the first series, this one diverges if x is between 0 and 1 but converges if x is greater than 1. Oliver was intrigued by the relationships between such pairs of series and investigated the mathematical and physical conditions under which either or both could be useful. He found that sometimes a *portion* of a divergent series could safely be used to compute numerical results which could otherwise be obtained only by much more laborious methods. But he insisted that whole series, including the 'unsafe' parts, should be considered as representing the physical processes.

All the time, he was driven by the desire to understand electrical circuits and fields. He hadn't gone into arcane areas of mathematics like fractional differentiation and divergent series for their own sake. He had done so because they turned up in his investigations of the physics and to have ignored them would have been a dereliction of duty. The way to make progress was to experiment boldly with the mathematics, guided all the while by the physics. As he put it:[13]

> It is by the gradual fitting together of the parts of a distinctly disconnected theory that we get to understand it, and by the revelation of its consistency. We may begin anywhere, and go over the ground in any way. Some ways will be preferable to others, of course, since they may be easier, or give broader and clearer views, but no strict course is necessary. It may be more interesting and instructive not to go by the shortest logical course from one point to another. It may be better to wander about, and be guided by circumstances in the choice of paths, and keep our eyes open to the side prospects, and vary the route later to obtain different views of the same country. Now it is plain enough when the question is that of guiding another over a well-known country, already well explored, that certain distinct routes may be followed with advantage. But it is somewhat different when it is a case of exploring a comparatively unknown region, containing trackless jungles, mountains and precipices. To attempt to follow a logical course from one point to another would then, perhaps, be absurd. You should keep your eyes and your mind open and be guided by circumstances. You have first to find out what there is to find out.

But the idea of experimental mathematics, throwing logic to the winds, was alien to the mathematicians of the Royal Society.

Not for the first time, Oliver was partly the author of his own misfortune. Had he made friendly overtures to the 'rigorists', acknowledging their essential role and going halfway to meet them, he might have had a response in kind. But diplomacy was quite outside Heaviside's range – he never went halfway to meet anybody. Scattered all through his work were passages that seemed designed to irritate orthodox mathematicians. He would toss in new ideas without defining them precisely, forge ahead with a method before submitting it to proof, and use apparently sloppy notation in a kind of shorthand. Faced with this 'take it or leave it' attitude, it is not altogether surprising that the Royal Society decided to leave it.

Oliver reacted to the rejection with a shrug of the shoulders rather than a shake of the fist. The disappointment was real and lasting but tempered by amusement. The Royal Society now had taken its place in his row of Aunt Sallys alongside archbishops and the House of Lords – pompous and antiquated objects of fun. He took the rejection without formal protest and later published a condensed version of Part III of his Royal Society paper in Chapter 8 of his treatise *Electromagnetic Theory*, Volume II.

The Society may have seemed to him rather ludicrous in its solemn formality, but to have the initials F.R.S. after one's name was still a mark of honour. Not that it cut much ice in South Devon society. Oliver recounted how a pleasant musical evening with family and friends at his brother's house had turned sour when the principal guest – a local bigwig – arrived. Everyone fussed around the newcomer, striving to make a favourable impression, and Oliver, who was actually the only person of distinction in the room, was ignored – apparently being thought not even worthy of introduction to Mr Big. There were attempts at apology after the guest had gone but they served only to heighten Oliver's scorn for conventional notions of social status.

One aspect of social status was, of course, money, and Oliver, in his mid-forties, was poorer than he had been as a young telegrapher with the Great Northern. He had hoped that his collected *Electrical Papers*, published in two volumes by Macmillan in 1892, might make his fortune but, despite having mostly good reviews on both sides of the Atlantic, the book had sold only a few hundred copies. The thought that a man of his genius should be so poorly recompensed for his work worried his friends and some of them persuaded the Royal Society to offer financial help. In February 1894, he had a letter from Fitzgerald, Lodge and another friend and admirer, John Perry.

> In further recognition of the acknowledged value of your contributions to science in carrying forward the work of Faraday and Maxwell, it has been proposed to grant you an honorarium by the Royal Society, inadequate though it will necessarily be.
>
> It has fallen to us as personal friends to have the honour, as well as the pleasure, of making this proposal known to you. As a matter of form and to avoid the humiliation of having to report a refusal, all we ask is an expression of your willingness to accept if offered, and thus give the greatest gratification to your friends and put them under a lasting obligation to yourself.

A generous offer, graciously presented by people he liked and trusted. But Oliver was obstinate in guarding his self-respect. He did not want to be given money out of compassion for his poverty and wanted to know the exact purpose of the fund from which the money would come. Fitzgerald didn't know and wrote to Lord Rayleigh, the Society's Secretary, to find out. Meanwhile, with saint-like patience, he replied to a volley of letters from Oliver and pleaded with him to see reason.[14]

> If you consider it humiliating not to be well off and not to enjoy sufficiently good health to earn more than a hodman, though the scientific world with one consent says that your work is worth that of hundreds of hodmen, I dare say I ought to say millions, then I am afraid that you will deprive the scientific world of the satisfaction of endeavouring to emphasize this, their belief, in a tangible way by paying you for a small part of the value they have received from you.

Oliver was not persuaded. His reply, by return of post, got to the heart of the matter: reward would be wonderful but charity was anathema.

> ... if the scientific world says my work is worth that of hundreds of hodmen or millions (I quote your words) and the scientific world chooses to pay me distinctly for that work according to their own (quoted) estimation, why then I should be proud to take it! It would be a pot of money.

However, even if I may have evidently emphasized in my previous a particular side of the question, still the fact remains that that side is a very real one. To give an extreme case, people sometimes starve rather than go to the workhouse. Why? The associations, I suppose, and then pride.

Then the requested information came from Lord Rayleigh. The money would be from the Society's 'Relief Fund', which had been set up 'for the aid of such Scientific men or their Families as from time to time require and deserve assistance'. That settled it. Oliver wrote again to Fitzgerald, refusing the offer with as much grace as he could summon up.[15]

Lord R is a man of few words. Or perhaps you have left out his opinion of my audacity, daring to look a gift horse in the mouth! Anyhow, it would not be seemly for me to ask more. Will you therefore be so good as to convey to Lord R my most respectful declination of the offer?

I fear I have been the cause of a great waste of your valuable time, all about nothing too. I am very sorry.

Ever constructive, Fitzgerald suggested that Oliver should try to earn some money by writing a popular book. The idea got short shrift. Oliver replied that he did not write 'for the masses' and, anyway, Volume 1 of his new treatise on *Electromagnetic Theory* had just come out and he expected it to sell much better than *Electrical Papers*. In any case, he added, 'if my income is small, so are my expenses, and I can scrape along'. So he could, thanks to occasional help from brothers Charles and Arthur, who had their own families to support and would have been only too glad to have had some of their extra burden relieved by the Royal Society.

A perverse and, perhaps, selfish decision by normal lights, but Oliver had at least held true to his own code. And he did have grounds for expecting better sales from the new book. For one thing, it had been planned as a book from the start, despite being written in instalments, whereas *Electrical Papers* was a retrospective collection. For another, he was by now a well-known figure in the scientific world, a sage consulted by people from far and wide. Among his illustrious new associates was the great American physicist Josiah Willard Gibbs, who became his ally in a new battle.

Eleven years older than Heaviside, Gibbs was Professor of Mathematical Physics at Yale University. Like many of the best nineteenth-century scientists, he covered a wide range of topics in his work. On electromagnetism he naturally turned to Maxwell's *Treatise* for enlightenment, as Heaviside had done, and was intrigued by Maxwell's use of *quaternion* notation to represent relationships between quantities such as electric and magnetic force which were vectors, having both strength and a direction in space. Quaternions had the advantage of great compactness and seemed to give clearer insight into the physical processes – orthodox representation needed many more equations and symbols, making it hard to see the wood for the trees.

Gibbs and Heaviside went through the same sequence. Both eagerly looked up the theory of quaternions, were disappointed to find it of little use, and went on to develop a simplified version that *was* useful. What they produced was the system of vector analysis, which eventually became the standard language used by physicists and engineers to describe fields. Astonishingly, the two men, working quite independently, developed systems that were identical apart from small differences in notation. Heaviside was by far the more prolific in using the new method, but where the notations differed it is Gibbs' version that has survived.[16] The honours are evenly balanced and there was never a dispute over priority: both were happy to share the credit. We have seen, in Chapter 5, how Heaviside's invention of vector analysis had helped him to turn Maxwell's wonderful but hitherto inaccessible theory into something physicists and engineers could use in their everyday work. He knew it was a historic innovation and one might have expected some disappointment at having to share the kudos. But Oliver was always generous with tributes where he felt they were due, and, in any case, to share credit with a man like Gibbs, sometimes called the American Maxwell, was an honour in itself.

Heaviside and Gibbs met opposition from many mathematicians who found the new methods – quaternions as well as vectors – strange and difficult, but the fiercest attack of all came from the small quaternion camp.

What were quaternions and why did they arouse such passion? They were the brainchild of the Irishman William Rowan Hamilton. Born in 1805, Hamilton was possibly the most gifted mathematician ever to come from the British Isles. As a young man he showed such brilliance that he was appointed Professor of Astronomy at Trinity College Dublin *before* he graduated. What is more, he hadn't even applied for the job. And it was a strong field – one of the candidates was a future Astronomer Royal of England, George Biddell Airy. Despite being hopeless at practical astronomy, Hamilton was then made Astronomer Royal of Ireland, simply to give him freedom to do mathematical research.

Starting with optics, he produced the greatest work ever written on the topic. His *Theory of Systems of Rays* brought together all previous results in a single theory so powerful that he was able to extend it to the whole of dynamics. It embodied the principle of least action and a new way of describing any mechanical system in a single function that expressed the relationship between its two basic quantities, energy and momentum. Hamilton's methods are now standard throughout physics and his name crops up everywhere because the characteristic function he defined for any system is called, simply, the hamiltonian. But he believed his greatest achievement was the invention of quaternions.

The idea started with the square root of minus 1. Mathematicians had discovered so-called complex numbers, which had the form $x + iy$, where x and y are ordinary numbers and i is the square root of minus 1. These strange numbers were proving remarkably useful and could be given a geometrical interpretation where i represented a rotation through 90 degrees – something like 'turn left'. Two such rotations, represented by i^2 or -1, meant 'turn left twice', or 'about turn'. By the same token, $-i$ meant 'turn right', which had the same effect as i^3, 'turn left three times'.

What Hamilton did was to take the idea, from complex numbers, of rotation in a plane, and extend it to three-dimensional space. In a plane there is only one possible axis of rotation about a given point but to specify a rotation in three-dimensional space three axes are needed, each at right angles to the others – we can think of one axis as vertical, allowing rotations such as turn left and turn right, and the other axes as north–south and east–west. So Hamilton needed three symbols, i, j and k, each to represent a rotation of 90 degrees about one of the axes. Repeat any of the rotations *i, j* or *k*, and you get an about turn. The problem of rotations had haunted Hamilton for fifteen years and the solution came to him in a flash while walking across the Brougham Bridge in Dublin. He was so thrilled that for a moment he forgot the dignity of his office of Astronomer Royal. He got out his pocket knife and carved the outline of his scheme on a stone of the bridge.

$$i^2 = j^2 = k^2 = ijk = -1$$

His symbols did not obey the traditional commutative law of multiplication – for example *ij* was equal to *minus ji* – but he found that with this modification the normal rules of algebra could be applied to super-complex numbers of the form $w+ix+jy+kz$, which he represented by a single symbol q. He called them quaternions and believed that they held the key to the mathematics of the physical universe.[17]

But Heaviside and Gibbs couldn't get the key to work. They both came to the view that Hamilton's generalisation of the complex number, beautiful as it was mathematically, did not bring any significant benefit to physics. How could it when a quaternion was the sum of two different things, an ordinary number, or scalar, *w*, and a vector, $ix + jy + kz$? Heaviside and Gibbs decided to simplify things by dropping the scalar parts, leaving pure vectors. Their thinking was reinforced by the fact that Maxwell had himself started the process of simplification when he identified the essential properties of *vector* fields, such as 'curl' and 'divergence', and gave them names.[18] Unlike quaternions, vectors *did* represent physical quantities, such as electric and magnetic forces, and did so in a way that enabled physical relationships to be described simply and vividly.

But for some people quaternions had acquired an almost religious status. Their leader was Maxwell's old school friend Peter Guthrie Tait. As a young man he had been enthralled by Hamilton's *Lectures on Quaternions* and the spell had kept its grip. When Hamilton died in 1865, Tait assumed the mantle and made the advancement of quaternions one of his main aims in life. A close friend of Thomson as well as Maxwell, Tait was an outstanding physicist in his own right, and a man of commanding presence, charming or fierce as the occasion demanded. He held strong views on just about everything and was never slow to take issue with anyone who saw the world differently. In fact, there was nothing he liked better than a good scrap. His wrath found many outlets but it brimmed over when Heaviside and Gibbs mutilated his sacred quaternions.

His first shot went in Gibbs' direction.[19]

> It is disappointing to find how little progress has recently been made with the development of Quaternions. … Even Prof. Willard Gibbs must be ranked as one of the retarders of Quaternion progress, in virtue of his pamphlet on *Vector Analysis*; a sort of hermaphrodite monster, compounded of the notations of Hamilton and Grassmann.

Gibbs rebutted the charge firmly but gracefully. Heaviside was less polite. He called Tait a 'consummately profound metaphysicomathematician' and added:[20]

> 'Quaternion' was, I think, defined by an American schoolgirl to be 'an ancient religious ceremony'. This was, however, a complete mistake. The ancients – unlike Prof. Tait – knew not, and did not worship Quaternions.

In response, Tait took a haughty tone: he could not bring himself to study the abhorrent methods of Heaviside and Gibbs in detail, but a colleague had done so and found it a grim experience.[21]

> Dr Knott has actually had the courage to read the pamphlets of Gibbs and Heaviside, and, after an arduous journey through these trackless jungles, has emerged a more resolute supporter of Quaternions than when he entered. ... Dr. Knott's paper is a complete exposure of the pretensions and defects of the (so-called) Vector Systems .... I find it difficult to decide whether the impression its revelations have left on me is that of mere amused disappointment, or of mingled astonishment and pity.

To illustrate the 'exposure', Tait gave a count of the number of symbols Heaviside had used in some of his equations (one contained 18 letters, nine suffices, three indices, three points, five signs and three pairs of parentheses), with the implication that such rough stuff did not bear comparison with compact and elegant quaternions. The comparison was unfair because it did not compare like with like, but Oliver let that pass. By taking such a snooty line, Tait had made himself an easy target. Oliver rubbed his hands and hit back:[22]

> The quaternionic calm and peace have been disturbed. There is confusion in the quaternionic citadel; alarms and excursions, and hurling of stones and pouring of boiling water upon the invading host. ... It would appear that Prof. Tait, being unable to bring his massive intellect to understand my vectors, or Gibbs', ... has delegated to Prof. Knott the task of examining them, apparently just upon the remote chance that there might possibly be something in them that was not utterly despicable. He has counted up the number of symbols in certain equations. Admirable critic!

Exchanges like these appeared in the journals for more than a year. They generated more heat than light but kept readers entertained. The Royal Society's Secretary, Lord Rayleigh, was too much of a gentleman to indulge in such antics himself but he enjoyed the spectacle from a ringside seat, observing wryly: 'Behold how these vectorists love one another'.

How Tait must have hated being called a 'vectorist'! He continued to fly the colours for quaternions but never rallied more than a handful of supporters. In the long run, Heaviside and Gibbs carried the day. Their vector analysis became first respectable, then indispensable. Today, vectors are everywhere and quaternions are rarely seen, though they are finding new uses in new areas such as computer graphics.

Scientific journals did not take sides in such debates but it happened that *Nature* carried most of the Tait/Knott contributions and *The Electrician* most of Heaviside's. Thanks to two outstanding editors, first Biggs, then Trotter, Oliver had had a tremendously fruitful partnership with *The Electrician* over the years. Then, in March 1895, he had a letter from Trotter.

> I regret to inform you that owing to serious interference with my editorial work by the secretary and the publishers, I am compelled to retire from The Electrician.

There were echoes from eight years before, when Biggs had been forced out, largely because of his support for Heaviside. This time the quarrel was about other things but the situation was still worrying: what were the chances of getting another editor as understanding and accommodating as Trotter? Oliver's luck held: the job went to assistant editor W.G. Bond, who proved to be as good a friend as Biggs and Trotter had been.

Bond soon found himself in a position familiar to his predecessors, caught between old adversaries Heaviside and Preece and exposed to fire from both sides. Preece, who only a few years before had wanted to banish inductance from transmission lines, had undergone a dramatic conversion. He now announced, as if it was his own idea, that a proper amount of inductance could be beneficial, and went as far as to propose a new type of undersea cable to harness this improvement and make it possible to link Britain and America by telephone. He intended to try it out on a link between the English mainland and the Isle of Wight. The design was ingenious: conductors with a semicircular cross-section would be bound with their flat sides together, separated only by a layer of insulating paper. But, somehow, Preece had got his calculations back to front. His cable would have had less inductance and more capacitance than a conventional cable – the worst possible result.

Asked by Bond to give an opinion, Oliver could see that this was not a time for heavy hitting. His old adversary had tied his own noose; anyone with a modicum of electrical theory could see that the proposal made no sense. He confined himself to a chuckle:[23]

> Mr Preece … thinks it right and proper to bring the two members of a pair of conducting leads as close together as possible. … Now, it is a long way to America, and one may never get there; but it is as well to go in the right direction.

When invited by Bond to respond, Preece dug himself deeper into trouble by insisting that his design did *not* increase capacitance – he had proved this by measuring the capacitance of a 300-foot test cable. But he gave no details of the experiment, nobody believed the result and he came in for a deal of mockery in the journals on both sides of the Atlantic. As for the new cable design, the Post Office quietly dropped it a few months later. But Preece had charisma in abundance and somehow, despite such gaffes, held on to his position as an international authority on electrical matters.

Oliver too, in his way, was becoming an authority, someone to be consulted on issues of the day, and he was asked to contribute to a debate on no less an issue than the age of the Earth.

Christian scholars had tried to calculate the date of the Creation from the scriptures. In 1642, Dr John Lightfoot, of Cambridge University, confidently gave the date as 23 October 4004 BC. He even specified the time of day – 9 am. Others came up with different conclusions but they all agreed that the Earth was less than ten thousand years old. The first serious challenge to a literal interpretation of the Book of Genesis came from the Scottish geologist James Hutton, who published his *Theory of the Earth* in 1795, claiming that our planet was very old indeed, with 'no vestige of a beginning, no prospect of an end'. Hutton was a genius but a poor advocate and his ideas made little headway until they were publicised and extended in the 1830s by the eloquent Charles Lyell. He attracted wide support for the theory that the Earth was in a continuous state of change brought about by natural forces – wind, rain, earthquakes, volcanoes and so on – and was, by evidence from rocks, many hundreds of millions of years old at the very least. Then, in 1859, came Charles Darwin's *Origin of Species*, with its theory that all life had evolved from the first simple organisms – a process that also required hundreds of millions of years.

Geology and biology both pointed to an almost unimaginably ancient Earth, but what did physics have to say? William Thomson took up the challenge in 1862, first with a rough estimate of the age of the Sun, then with some more complex calculations of the age of the Earth. He reasoned that both bodies were cooling, so by knowing their current states and making reasonable assumptions about their original states he could use the theory of heat flow to work out how much time had elapsed. He came up with ages of around 100 million years for the Sun and 98 million years for the Earth – much older than the Biblical estimates but not nearly old enough for the theories of Darwin and Lyell.

Science was in a pickle. Somebody had to be wrong but nothing can defy the laws of physics and it was almost unthinkable that the great Thomson could have made a mistake. On the other hand, Lyell's theory had what seemed to be irrefutable evidence in its favour and many people thought the same of Darwin's. Thirty-three years passed before anyone managed to make an effective assault on the impasse. The assailant was Heaviside's friend John Perry, who had been Thomson's assistant at Glasgow University and was now Professor of Mechanical Engineering at Finsbury Technical College.

Perry thought that the geologists must be right, that the evolutionists were probably also right, and that Thomson must therefore have gone wrong somewhere. The weakest part of Thomson's analysis was his assumption that all the material making up the Earth had the same propensity to diffuse heat – that the *diffusivity* was the same at all depths below the surface. Perry thought it much more likely that diffusivity increased with depth and he reworked the calculations accordingly. Making reasonable assumptions about the Earth's composition, he came up with a new range of possible values for its age. The lowest was about 300 million years and the highest about 30 *billion*. He boldly gave the opinion that Thomson had seriously underestimated the age of the Earth.

By changing the assumption on diffusivity, Perry had given himself a frighteningly difficult mathematical task and he tackled it in two ways. The first was to make some ingenious simplifications and do the calculations himself; the second was to ask his friend Oliver to check the answers. Things couldn't have worked out better. He happily reported:[24]

> Mr. Heaviside has given exact solutions, and has found that there is practically no difference between mine and the exact numerical answers. That Mr. Heaviside should have been able, in his letters to me during eleven days, to work out so many problems, all seemingly beyond the highest mathematical analysis, is surely a triumph for his new working methods.

Good publicity for Heaviside's operational calculus, which seemed to have almost supernatural power. But to Oliver, it was Perry who had worked the magic by coming so close to the exact results by rough and ready methods. A notebook entry asks, with admiration: 'Is he a witch, or is this one of the most remarkable coincidences in ancient or modern history?'

Thanks to Perry, physics no longer flatly contradicted geology and biology, but his reconciliation was illusory. All later became clear when hitherto unsuspected sources of heat energy were discovered – radioactive decay in the Earth and thermonuclear reactions in the Sun. Thomson's and Perry's calculations of the Earth's cooling were rendered irrelevant. But they hold historic interest and, perhaps, a lesson. When faced with a similar impasse in the kinetic theory of gases, Maxwell had said the only thing to do was adopt the attitude of 'thoroughly conscious ignorance that is the prelude to every real advance in science'. In his case the advance turned out to be the discovery of the quantum of energy, though he didn't live to see it.

What of Heaviside's part in the controversy? For once, he was not a combatant but an 'authority' called in as an expert witness, and he loved it. He kept his distance from the battlefield and was sceptical about the value of Thomson's and Perry's results. He never commented formally on Darwin's theory but went along with it and summed up his views on evolution very nicely in a letter to Fitzgerald:[25] 'Wonderful how things have worked out. If it wasn't true, no one could believe it.'

Meanwhile, life in Paignton had changed. The supposed health-giving powers of the Torbay climate had failed to work their magic on Oliver's parents, Thomas and Rachel. The move from smoky London, where they had lived for over 50 years, had brightened their lives but they were by now worn out. There were good days – including a memorable family outing to Berry Pomeroy Castle on Thomas's 80th birthday, for which Arthur and family came down from Newcastle – but they both grew increasingly frail. As resident nurse, Oliver had many sleep-deprived nights. Unlike his hero Maxwell, he had no natural gift for tending the sick, but he did his best, and was often bone-tired from the effort.

Rachel died in October 1894 and, at about the same time, Oliver had sad news from Germany. Heinrich Hertz had died from an infection of the jaw at the tragically early age of thirty-six. The two men, who could not have been more different in upbringing and temperament, shared a passion for their science and had built up a firm friendship through their letters. During his short visit to London, Hertz had only narrowly been dissuaded from making a special 400-mile round-trip to Paignton. Oliver had said he was 'shocked' that Hertz should even have thought of such a thing but, of course, neither of them knew that it was to be their only chance to meet.

When Oliver refused financial aid from the Royal Society his friends did not give up trying to help him. Led by John Perry, they doubled their efforts and in 1896 managed to persuade the Government to offer him a pension of £120 a year from the Civil List. This was an entirely different proposition from the Royal Society's offer, a reward from a grateful country with no whiff of charity about it, and Oliver was delighted to accept. What was more, the pension was backdated for nine months, so he had £90 in his pocket.

A man of means? Not exactly. Over the years the loans from his brothers had accumulated, so there were debts to pay. But more help soon arrived: he had a package from the London Joint Stock Bank containing £100 in banknotes. It came, according to the covering letter, 'at the request of Mr John Perry' – presumably Perry had organised a collection among his scientific and engineering colleagues. We don't know whether the money was a gift or a loan, nor whether Perry had consulted Oliver about it beforehand. But we can be sure that Oliver didn't send it straight back. A note he made on the back of the letter shows that, after recording the serial numbers of the notes, he posted £45 to Arthur the same day and gave £15 to Charles.

Perhaps it was the exhilaration of having cleared his family debts that gave him, for a while, a remarkably optimistic outlook on life. When his father Thomas died in November 1896 and he was left alone in the flat over the music shop Oliver decided, for the first time in his life, to take charge of his own domestic arrangements. He looked for a house of his own and, after a long search, found one – a 'gentleman's house' at Newton Abbot, a few miles inland, with the grand name Bradley View. In June 1897 he struck out to begin a new life as a country gentleman.

Chapter 10
Country life
Newton Abbot 1897–1908

It was the boldest experiment of his life. Helped only by a housekeeper, whom he brought from Paignton, he undertook the complete organisation of his domestic life. Oliver was ill-equipped for such a venture. The hundreds of little interactions with other people that make up day-to-day life were to him a trial and a burden. It wasn't simply laziness or selfishness, although that was the way it must have looked. His system for navigating life's hazards seemed to work differently from everybody else's: his antennae picked up different signals and his brain interpreted them in a way that made little allowance for human frailty other than his own.

He began the adventure in high spirits and wrote straightaway to tell Fitzgerald about the move.[1]

> Behold a transformation! The man 'Ollie' of Paignton, who lives in the garret in the music shop, is transformed into Mr. Heaviside the gentleman who has taken Bradley View, Newton Abbot, that disagreeable residence which has been empty so long. Built for a gentleman's house (small) it was occupied for very long by a farmer, and got into a very disreputable condition in consequence. But 30 pounds or so has been spent on it, mostly inside, and it is now not so bad. It is such a change! Very pretty place. Fields gloriously arrayed with buttercups and fringed with proper hedges with plenty of trees (it would shock a Scotch farmer) come right up to the garden, and a wood and hills behind them make up the picture.

One of his first visitors was G.F.C. Searle, a young physicist from Cambridge who had already called to see him several times in Paignton. Searle lived until the 1950s and in later life prided himself as one of the few people still alive to have met both Maxwell and Heaviside. As a boy he had been taken by his father to the Cavendish Laboratory as a birthday treat. They were given a conducted tour by Maxwell himself, and the great man spent an hour showing them experiments. It was enough to inspire a lifetime's dedication to science. Searle is remembered by people still alive today as a fierce disciplinarian in the lecture room and the laboratory, but in his younger days there were light-hearted moments and some of them were spent in South Devon.

Searle had written to point out an error in one of Oliver's papers and had received such a friendly reply that he was emboldened to call in while on a cycling holiday. It was the first of many visits. The two men grew very fond of each other and Searle came closer than anyone else to understanding his friend's obdurate ways. An incident from one of his early visits to Bradley View serves to demonstrate this.

On the way to Newton Abbot, Searle had called at Torquay to see Charles Heaviside, who said that Oliver never came to visit them, though they would have liked him to. The next day Searle and Oliver decided to go out on their bicycles. Know-ing better than to suggest a visit to Charles, Searle said 'let's go to Babbacombe', knowing that the road ran near Charles' shop. As they were approaching Torquay he said 'We had better see about getting something to eat.' As if on cue, Oliver said 'I'll tell you what we'll do. We'll go and see my brother; he'll give us lunch.'

They explored many miles of the Devon lanes on bicycle rides, often 'scorching' down steep, rough and twisty lanes, with feet up on the front forks out of the way of the whirling pedals. Oliver would fold his arms, steer by leaning the bicycle over, and leave the more cautious Searle far behind. Such recklessness didn't go unpunished: he had a nasty spill when a chicken ran across the road and there was at least one other occasion when his bicycle had to go in for repairs.

George Francis Fitzgerald was another enthusiast for bicycling. He had had his share of accidents, too, including one with a chicken. In one letter to Oliver he sketched a radical new design, in which the rider reclined almost horizontally and had the luxury of pneumatically sprung handlebars, but admitted that he hadn't yet worked out how to make it steer. He paid Oliver a memorable visit in 1898. They rode out together but Fitzgerald could stay only for the day because he had to return to Dublin.

The three men never met all together but their interaction played a preliminary part in one of the great triumphs of twentieth-century physics – the theory of rel-ativity. They were all intrigued by the question: how does the shape of the field around an electric charge change when the charge moves? This was the very topic on which Searle had originally written to Oliver – by applying Maxwell's theory, they had predicted that when the charge approached the speed of light the shape of its field would become compressed along its direction of motion, for example the field around a point charge would change from spherical to ellipsoidal.[2] Fitzgerald took the idea further and proposed that the same held for all matter – that *all material bodies* contracted along their direction of motion as they approached the speed of light. If true, this hypothesis would explain the strange result of the experiment of Michel-son and Morley.[3] They had found that light always appeared to travel at the same speed, no matter how fast or in what direction the observer was moving. According to Fitzgerald, this happened because the observer's instruments (and the observer himself) contracted by exactly the amount needed to compensate for his movement.

Fitzgerald's idea seemed crazy but, with the interpretation later put on it by Einstein in his *Special Theory of Relativity*, it is now taken as a fact of nature.[4] The Dutchman Hendrik Anton Lorentz had the same idea independently and both are honoured in the title: the Lorentz–Fitzgerald contraction. A simple formula comes with it. The factor by which a body appears to contract as it approaches the speed of light is the same as that by which a clock on the body appears to slow down. It has the value $\sqrt{(1 - v^2/c^2)}$, where v is the velocity of the body and c is the speed of light. It occurs throughout special relativity and is sometimes called simply the relativistic factor. It is also exactly the factor identified by Heaviside and Searle in their work on the electric field of a moving charge.

Oliver had few visitors besides Searle and Fitzgerald, and the citizens of Newton Abbot became curious about the solitary newcomer. Naturally so, but Oliver was unschooled in the ways of country people. He took their gaucherie for rudeness and their well-meant enquiries for prying. He had tried very hard at first to get on with his neighbours but couldn't sustain the effort, and a vicious circle set in. He became more and more suspicious of the locals and pulled further into his shell; they regarded him more and more as a misfit and, eventually, as an object of ridicule. A letter he wrote to Fitzgerald shortly after his friend's visit, and just over a year after moving in, shows the way things were going.[5]

> In spite of the secretive manner in which you entered and left this town, you did not escape notice, either on your arrival at the railway bridge in the morning, or your departure up the hill in the evening. 'Who did the ... go out with? Was it his father? Was that the old man's bike standing outside?' Certainly the rudest lot of impertinent, prying people that I ever had the misfortune to live near. They talk the language of the sewer, and seem to glory in it. You would be astonished if I were to go into detail about the way they have baited me. But I have not the least doubt of some hanky panky behind it.

He never came to terms with the normal way of life in Devon towns, where everyone's business was everyone else's and where – as a West Country colleague once told me – the expression 'daft old bugger' was a common term of endearment.

By a strange irony, just as Oliver's isolation from his local community grew, so did his national and international reputation. The little house, its tenant scorned by the locals, was, according to one writer, regarded by many men of science as 'a temple of wisdom, the place of the oracle, the court of ultimate appeal'.[6]

People wrote to ask Oliver's opinion on all kinds of topics: seismometers, the choice of site for the new National Physical Laboratory, and one that was the talk of the day – Wilhelm Konrad Röntgen's discovery of X-rays. To the public, the idea of being able to see through solid objects seemed like science fiction come true and the press had a great time with jokey scare stories. For example, *Punch* suggested that underclothes should be made from lead to preserve people's modesty. For physicists there was one main question: were the vibrations in the new rays transverse, as in light waves, or longitudinal, as in sound waves? Oliver's strong views on this led to a dispute with the illustrious Austrian physicist Ludwig Boltzmann. Both were ardent supporters of Maxwell but Oliver contended that Boltzmann was perverting Maxwell's theory by interpreting it in a way that would allow longitudinal waves to exist as well as transverse ones. The new X-rays, Boltzmann thought, could be *longitudinal* Maxwellian waves. Oliver brooked no such nonsense, writing:[7]

> There are no 'longitudinal' waves in Maxwell's theory analogous to sound waves. Maxwell took care that there should not be any.

He was right. The mysterious X-rays turned out to be just like light waves, but with a much shorter wavelength. And both were parts of a huge spectrum of Maxwellian waves with wavelengths ranging from nanometres to kilometres – all of them transverse.

Oliver was on home ground against Boltzmann, but he strayed into unfamiliar territory when drawn into a dispute between the President of the Institution of Electrical Engineers, James Swinburne, and his predecessor but one, John Perry. They were both admirers of Heaviside but agreed on little else and got into a bitter and public argument about, of all things, the second law of thermodynamics. Swinburne was confident and wanted an emphatic victory, so he appealed to the highest authority on the topic, Max Planck, for a judgement. Success! Planck said that Swinburne had produced 'one of the best and clearest expositions of the subject that has ever been written'. Praise indeed, but Oliver decided to challenge Planck. Whether he did so in support of Perry or out of sheer devilment we can only guess, but it was a hazardous venture, like a weekend golfer taking on Tiger Woods. As usual, he struck boldly, writing in *The Electrician*:[8]

> I should like to ask Professor Max Planck whether the view he expresses that 'Nature never undertakes any change unless her interests are served by an increase in entropy*' is to be taken with or without any particular reservation or with any special interpretation of 'her interests'. My thermodynamical ideas are somewhat old-fashioned – viz., that there is invariably a dissipation of energy or loss of availability of energy due to imperfect or total want of reversibility in natural processes. This entirely agrees in effect with the way of expressing things in terms of 'entropy', although that subtle quantity is certainly 'ghostly', and is somewhat too evasive to be regarded as a physical state even though it be a function of the physical state referred to a standard state. But the question is how the interests of Nature are served by imperfect reversibility? Professor Planck's words suggest a choice on Nature's part, as if Nature had any choice. Goethe said God Himself could not alter the course of Nature. That was truly scientific. Then, again, what are to be considered the interests of Nature? Are we to take things exactly as we find them, and define the interests in that way? If so, it carries us no further. Or is there a theorem of greatest entropy, showing how any variation from the proper course of Nature would tend to reduce the rate of increase of the entropy?

Entropy was the mysterious quantity that always tended to increase and thereby ensured, among other things, that heat flowed from hot bodies to cold rather than vice versa. To Heaviside, as a casual visitor to the subject, it looked like a mathematical contrivance, useful for calculation but not a fundamental property of matter. Planck disagreed, and there was a touch of exasperation in his reply – rash dabblers should refrain from commenting on things they don't understand.[9]

> Whether entropy has any 'ghostly' attributes is a question I will not open, but I am for the present quite content to know that it is a quantity which can be measured without ambiguity. … I do emphatically deny, and have always combated the proposition adduced by Mr. Heaviside, of the universal dissipation of energy.

This was enough to show Oliver that he had bitten off more than he could chew. For once, discretion prevailed, and he withdrew from the fray. To Planck, Heaviside's observation that all natural processes involve waste of energy was a clumsy oversimplification. For example, the energy lost by atoms when they emitted electromagnetic radiation could be completely recovered by reabsorption – none was wasted. It was the study of this process in so-called 'black body' radiation that led Planck to the discovery of the quantum of energy. But the discovery came only after Planck had discarded his earlier ideas about entropy and accepted that it did have a

---

* Entropy is a measure of the disorder in a system. According to the second law of thermodynamics, the total entropy of a closed system always tends to increase. A small change in entropy is defined as the amount of heat transferred in a reversible process divided by the temperature at which the heat is transferred.

certain 'ghostliness' – it was not a physical property of matter but a *statistical* one. By the same token, the second law of thermodynamics was a *statistical* law. It was physically possible for heat to flow from a cold body to a hot one: the only reason it didn't happen was that the probability of such an outcome was tiny. The startlingly original statistical ideas to which Planck became a reluctant convert came from his compatriot Boltzmann and – who else? – good old Maxwell.

If entropy was a measure of disorder in a system it was certainly on the rise in the Heaviside household. Oliver's housekeeper, whom he had brought from Paignton, suffered a stroke and never properly recovered. He described what followed in a letter to Lodge.[10]

> Poor woman sent away. No good for hard work again, I fear. Then had a charwoman two days. She left off coming, since when I have been alone. Quite independent, and have whatever I like for dinner. Stone broth, ditchwater soup. Made several discoveries. Parsnips cook easily. Carrots don't. So if you boil them together, the same time, when the parsnips are done, the carrots are as hard as stones; and when the carrots are done, the parsnips have lost all the flavour of proper parsnips. You mustn't pour anything hot into a glass dish. Catastrophe. Bang goes sixpence! If a pound of beef is used to make soup, and is kept boiling day after day, how long will it take to disappear? Haven't found out yet. …
>
> I have adopted the Principle of Least Action,[†] a most clumsy machine in electromagnetics, but is splendid in the house; assisted by the older principle that Prevention is better than Cure. E.g., nasty job blacking boots. Don't black 'em; use tan boots. Fires is a most horrid nuisance, with the dirt and the work. Abolish them. Use gas fires; no more trouble and labour. I have four, a gas cooker in the kitchen and gas fires in the sitting room and bedroom. It is such a blessing, that I am always thinking how to get gas or something to do the rest of the housework.

Household cleaning took the lowest priority and heating the highest. The mysterious 'hot and cold' disease still plagued him and to help fend it off he kept his fires going full pelt in the winter; he probably had the highest domestic gas bill in the whole of Devon. Visitors found the house as hot as a Turkish bath but Oliver would still be sitting by the fire with a dressing gown over his ordinary clothes and an eiderdown over his legs. Searle's visits were always welcome, especially when he brought his new wife, and Oliver did his best to play the congenial host. Searle reports:

> We went to Newton Abbot and Mrs. Searle took him some flowers. We had tea with him. We had been warned as to what we should find. The tea-pot spout was completely stopped up by tea leaves and no tea could come out of it. Oliver tipped the pot so far that the tea ran out of the top. He caught what he could in the cups, and carefully spooned the tea leaves out of Mrs. Searle's cup.

When inviting the Searles next time, Oliver wrote on the back of the envelope 'No flowers by request'. Visits were usually at tea time and Oliver's technique improved a little with practice. Searle reports again:

> After some conversation he would say 'Now I must go and get the tea ready' and you would hear him cleaning the knives, and then throwing them down. His habit was to throw things rather than to put them down quietly. Then he would ring a little bell and we went into tea.

[†] A fundamental principle in dynamics, formulated by Pierre Louis Moreau de Maupertuis in 1746 and later developed by William Rowan Hamilton and others.

He said one day 'There are nine pieces of bread and butter – three pieces each. There is some cake at the end but I don't recommend it.'

Oliver seems to have given up having overnight guests soon after his housekeeper left. The Searles often stayed in holiday accommodation in Torquay for a week or two, paying Oliver several visits each time, and sometimes acted as intermediaries between the two brothers. Charles was an amiable soul who chatted to everyone. He got on well with the people of Newton Abbot, as he did with everyone else, and assured Searle time and again that Oliver's complaints of harassment by locals there were groundless. As we'll see, he wasn't always right, but he did what he could to help his maverick brother. Knowing that Oliver had greatly missed music since leaving Paignton, Charles sent him one of the new mechanical pianos that had just started arriving at the shop. They came in slightly different forms under various trade names, one of which, the Pianola, became generic, like Hoover. The machine 'read' the music in the form of perforations in a roll of paper and played it for you. All the 'pianist' had to do was to work some bellows with a foot pump and adjust a hand-turned knob to control the volume. This was just the thing for Oliver, but had its drawbacks, as one of his visitors recounted:[11]

> I remember how he used to use the pianola with great vigour, so that, notwithstanding his deafness, he could hear the music to his own satisfaction, but to the discomfort (not knowingly) of other persons in the same room.

The pianola was the creation of an enterprising American, Edwin S. Votey, who had built a prototype in his home workshop in Detroit and persuaded the Aeolian music company to take it up. At the time it arrived in Britain, other American inventors were working hard on Heaviside's idea for ridding telephone lines of distortion. They were ambitious men with a simple motive – to make money. In the early days of the telegraph the same spirit of commercial enterprise had prevailed in Britain, and men like Cooke and Wheatstone, even Thomson, had made their fortunes. But since the Post Office secured its monopoly in 1870, the buzz of invention had faded. William Preece and his colleagues protested otherwise, but state control had stifled progress. Ambrose Fleming, who later transformed telecommunications by inventing the thermionic valve, felt so strongly on the matter that he published an essay called 'Official Obstruction of Electrical Progress'. He claimed that when the telephone first came to Britain from America Preece and his Post Office colleagues had 'laughed at it as a toy', and continued:[12]

> It is a pure waste of time for an inventor to spend days and nights over a telegraphic[‡] invention, or invest capital in patenting it, unless he can get it tried, and, if it succeeds, market his invention to a purchaser. He is not generally a philanthropist, but is spurred to work by the hope of reward. But in electric telegraphy he can try nothing and market nothing unless he first persuades the permanent officials of the State Telegraph Department. He has to overcome their inertia, opposition, or it may be ill will, before he can even get a trial of his telegraphic apparatus, and when at last he demonstrates an important advance,

---

‡ For 'telegraphic', one can read 'telegraphic or telephonic'. In legal terms the telephone was a form of telegraph.

he is entirely at their mercy whether it shall be adopted or not, and, if so, what price he shall receive for it.

We can see why Oliver saw no point in trying to get the Post Office interested in his recipe for a distortion-free line. Someone else did try. Sylvanus P. Thompson put up a proposal to fit high-inductance coils at intervals across the line, like ladder-rungs, but didn't get far. His scheme would not have worked, anyway. As Heaviside told him, he had put the coils in the wrong place – they should have been put directly into the line rather than across it.

Things were different in the Land of the Free. Engineers there had been the first to tackle the problem of cross-talk on telephone lines by bringing in so-called metallic circuits. These were circuits that used metal conductors for both outward and return currents, rather than the single outward conductor and earth return which had been standard for telegraph work. They almost eliminated cross-talk but the even greater problem of distortion remained and the best inventors were in a race to crack it.

One of the front-runners was George Ashley Campbell. A keen follower of Heaviside's papers, he had studied at both Harvard and the Massachusetts Institute of Technology before being hired by the American Telephone and Telegraph Company to work at their research department in Boston. He knew from Heaviside's work on the transmission line that distortion could, in theory, be eliminated by distributing the right amount of inductance continuously along the line, and that one could get close to this ideal by inserting coils of high inductance and low resistance at equal intervals. The remaining problems were practical ones. Could the coils be made with sufficiently low resistance? And how far apart should they be spaced? Spacing would have to be a compromise – many coils, closely spaced, would spread the inductance evenly but would be very expensive. Benefit had to be weighed against cost and the key commercial question was: what was the maximum spacing of coils that would still give a worthwhile reduction in distortion?

Campbell came up with a design to fill the bill – coils that would not be too expensive to make, spaced at around one-mile intervals. Tests went well and he filed a patent application. But just when the world seemed to be at his feet, a rival appeared. Professor Michael Pupin of Columbia University in New York had come to similar conclusions about how to improve telephone lines and filed a competing claim.

Pupin was a living embodiment of the American Dream. He came from Serbian peasant stock – his parents could not read or write – and had arrived in America as a boy of fifteen, alone and with five cents in his pocket, determined to make his fortune. Starting as a farmhand in Delaware, he worked his way to college and a quarter of a century later was Professor of Engineering at Columbia with plenty of ambition still waiting to be fulfilled.

What about Heaviside's rights in the matter? Not only had he published the formula for a distortion-free line in 1887, in 1893 he had suggested that the distortionless condition could be approximated by inserting coils of high inductance along the line.[13]

> Instead of trying to get large uniformly spread inductance, try to get a large average inductance. ... This means the insertion of inductance coils at intervals in the main circuit. That is to say, just as the effect of uniform leakage may be imitated by leakage concentrated

at distinct points, so we should try to imitate the inertial effect of uniform inductance by concentrating the inductance at distinct points. The more the better, of course.

Brief and clear. The idea of inserting coils in the line was demonstrably his, even though he had not taken out a patent. But this suggestion was hidden away near the end of his long series of papers on the 'Theory of Plane Electromagnetic Waves'. In the context of his great project at the time – to present to the world the true theory of electrical communications, based on Maxwell's principles – this was exactly the place for it. But perhaps he should have displayed the idea more prominently, or gone into more detail on, for example, the coil spacing. Then the outcome of the patent contest might have been different, bringing him at least public acclaim, if not fortune.

The Patent Office took its time coming to a decision. At first both applications were rejected on the grounds that Heaviside had already published the invention, but Campbell and Pupin contested the finding – Heaviside may have published the general principle but he had not taken out a patent and in any case they had taken things to a further stage with their specific designs. It was a tense time; the top men at the American Telephone and Telegraph Company were nervous and their fears were borne out when, in June 1900, Pupin won his case. It was a narrow decision and they could have contested it in the courts: after all, their man Campbell's system had been built and tested whereas Pupin's existed only on paper and in a laboratory simulation. If they lost they could still secure a monopoly on the design by paying Pupin for the rights. But what if the court went back to the first finding and negated all patents because they were based on Heaviside's writings? There would then be a free-for-all, with no chance of getting a monopoly. AT and T decided to play safe. They abandoned Campbell and bought Pupin out.

Pupin had made his fortune. The company paid him $185,000 on the spot and $15,000 each year during the seventeen-year life of the patent. His talent for self-promotion came into its own and the patent brought him not only wealth but fame. Every coil inserted into a telephone line was a 'Pupin' coil and the process became known as 'pupinisation'. He played a constructive part in the scientific establishment of his adopted country, for example helping to found the National Research Council. But perhaps his most remarkable achievement was in literature. In 1924 he won a Pulitzer Prize for his autobiography *From Immigrant to Inventor*. It is indeed a page-turner, but reads too much like a Hollywood rags-to-riches story to be entirely credible. He drops famous names freely but scarcely mentions Heaviside and doesn't mention Campbell at all. The only people he credits with helping him to invent the inductively loaded transmission line are the Serbian herdsmen who had shown him how to send sound signals by tapping on the ground.

Inductive loading freed the telephone from its shackles. Within a few years you could make a call from New York to Denver, 2700 miles away. AT and T made such huge profits that their payments to Pupin became insignificant but, as Pupin himself points out in his book, the real beneficiaries were the ordinary people who used the telephone. Campbell felt he had been swindled and, although he went on working for AT and T and collected many successful patents on the way, he carried the resentment for the rest of his life.

So did Heaviside. Nice though it would have been to have had money, the fact that it passed him by did not trouble him greatly. What really hurt was that his role as the principal author of this great technical advance went unrecognised because Pupin ran off with all the glory. To have the credit for his own achievement attributed to someone else was bad enough but to have it stolen by a rogue was worse. And Oliver was in no doubt that Pupin was a rogue; he had condemned himself. Early on, Pupin had written glowingly of Heaviside:[14]

> Mr Oliver Heaviside, of England, to whose profound researches most of the existing mathematical theory of electric wave propagation is due, was the originator and most ardent advocate of wave conductors of high inductance. His counsel did not seem to prevail as much as it should, certainly not in his own country.

Yet he was quick to drop any mention of Heaviside as an 'originator' as soon as it became clear that to do so might jeopardise his own claim to fame.

Oliver never stopped trying to dislodge the usurper from the seat of honour but it was like trying to turn back the tide. An early opportunity came when he was asked to write an article on 'The theory of electric telegraphy' for the 10th edition of the *Encyclopædia Britannica*. Having asserted his claim to the *principle* of distortion-free transmission – spread the right amount of inductance uniformly along the line – Oliver continued:[15]

> The writer invented a way of carrying out the principle other than uniformly, and recommended it for trial; viz., by the insertion of inductance coils in the main circuit at regular intervals .... In America, some progress has been made by Dr. Pupin, who has described an experiment ...

Neatly done. In a few crisp words he had stated his own priority claim and damned Pupin with faint praise, giving him a pat on the head for good measure. And the words were now enshrined in that esteemed repository of knowledge, the *Encyclopædia Britannica*. But they made no headlines and the Pupin bandwagon rolled on. By an extraordinary irony, Oliver had spiked his own guns by giving the popular press a far more newsworthy item in the same article. This was a prediction that made him for a while almost a household name, and for which he is most widely remembered today even though it comes well down the list of his achievements – the existence of the Heaviside layer in the upper atmosphere.

The scope of the *Britannica* article took in wireless telegraphy, including Guglielmo Marconi's amazing achievement in sending signals across the Atlantic from Cornwall to Newfoundland. Marconi knew little theory and refused to be put off his experiment by experts who said that electromagnetic waves travelled in straight lines and would never bend round the Earth's surface. When he confounded them, they were forced to seek an explanation. Why didn't the waves simply go off into space? Some scientists, including Marconi's own favoured adviser on such matters, Ambrose Fleming, thought that the atmosphere must be acting as a sort of lens, bending the waves by refraction. Heaviside thought differently. In ordinary telegraphy and telephony waves didn't travel *in* a conducting wire, they travelled alongside it – they were *guided* through space by the wire. The material of the Earth was a conductor – seawater being a very good conductor indeed – and might itself act to guide the waves round the Earth. This was not all. He announced:

> There is another consideration. There may possibly be a sufficiently conducting layer in the upper air. If so, the waves will, so to speak, catch on to it, more or less. The guidance will be by the sea on one side and the upper layer on the other.

The best part of a century later, packed audiences at the stage musical *Cats* were entranced by the scene where Grizabella, the fallen glamour cat, having sung her big number 'Memory', ascends, radiant and redeemed, to the Heaviside layer. Oliver had given T.S. Eliot, on whose poems *Cats* is based, the perfect name for cat heaven.

In fact, Oliver's name was usually coupled with that of Arthur Kennelly, an expatriate Briton working in America, who had independently made a similar prediction a few months earlier. In the 1920s, Appleton and Barnett found that a reflective layer really did exist about 60 miles above the earth's surface and further investigations showed that it was part of a complex ionosphere. It now goes by the prosaic title 'the E layer'. T.S. Eliot would not have approved.

The propensity of wireless waves to follow the curvature of the Earth was remarkable for another reason: it was one of the rare topics on which Oliver Heaviside and William Preece agreed. To his credit, Preece had been one of the few 'experts' to encourage Marconi to try sending signals across the Atlantic. He had also been instrumental in getting the young Italian started when he first came to Britain in 1896 seeking help for his experiments. Their ways parted a few months later when the astute Marconi began to feel the bonds of patronage tightening and cut himself free.

Early in 1901, Oliver had a letter from Lodge with worrying news from Dublin – Fitzgerald was ill. He had suffered for many years from digestive problems but until now had bounced back vigorously from each attack. Not this time. The man of teeming ideas was under strict orders 'not to think at all', and Lodge had written round to all their mutual friends warning them to send no letters so that Fitzgerald would have a chance to recover. But he never did. A last-hope operation failed to save him and he died aged forty-nine leaving a widow and eight young children. A gregarious, generous-spirited and self-effacing man, Fitzgerald had been a tremendous source of ideas and encouragement to many of his contemporaries. His loss was keenly felt across the whole scientific community – perhaps most of all by his fellow Maxwellians Lodge and Heaviside. In a heartfelt obituary, Lodge said that Fitzgerald was always 'the life and soul of debate' and that he 'loved him as a brother'. Heaviside sent his own commiserations to Lodge.[16]

> I understand and sympathise with your grief, for knowing him so much better than myself. I only saw him twice but we had a lot of correspondence at the time and I came to love the man. There was a considerable mutual understanding, to say nothing of his kindness to me.

He wanted to write to Mrs Fitzgerald but didn't have her address and, at Lodge's suggestion, wrote to Fitzgerald's colleague John Joly asking for it. For some reason Joly didn't reply, and Oliver had to get Lodge's help in wresting the address from him. At length it came, with a curt note: 'Sir O. Lodge instructs me you desire Mrs F's address. Here it is.' Joly went on Oliver's black list, but Mrs Fitzgerald was glad to have his condolences and replied straight away with a charming letter – a 'regular woman's letter', as he told Searle, 'Pride; and joy; not forgotten; loss hard to bear, etc.'

When the time came to compose a dedication for Volume III of his magnum opus *Electromagnetic Theory* Heaviside wrote:

In memory of

GEORGE FRANCIS FITZGERALD

'We needs must love the highest when we know him.'

Fitzgerald was indeed the 'life and soul of debate'. He and Heaviside had many friendly but spirited discussions and didn't always manage to resolve their differences. Fitzgerald had been trained in orthodox methods and tended to take a more conservative stance than the radical Heaviside. A case in point was one of Heaviside's great causes – a campaign to make a fundamental change to the system of units in which electrical and magnetic quantities were measured. Fitzgerald tried time and again to get Oliver to drop his campaign, not because he thought it was wrong but because he thought it was futile – that Oliver was simply wasting his time trying to overturn the massive vested interests that shored up the existing system.

It was all to do with the numerical factor 4π that cropped up nearly everywhere in the formulae of electricity and magnetism. Heaviside hated the 4π s – he thought they had no business there and wanted to get rid of them. His argument went to the very origins of the science of electricity.

Charles Augustin de Coulomb had discovered in the late eighteenth century that the force between two electric charges obeyed an inverse square law: the force between two charges in empty space (or air) was proportional to their product divided by the square of the distance between them. The constant of proportionality was originally taken to be 1 but when physicists and engineers found that this gave an impracticably tiny unit of electric charge they changed it to $1/\varepsilon$, where ε was a very small number that came to be called the primary electric constant[§]. So the formula for the force F between two charges, each of value q, separated by a distance r was:

$$F = q^2/\varepsilon r^2$$

So far so good. But as physicists developed the theory of electricity and magnetism many of their equations turned out to contain the numerical factor 4π, sometimes in the most curious places. As the theory advanced, more and more equations came along, complete with 4π s. Theoretical physicists got used to the clutter but the effect on newcomers was bewildering. Heaviside called the whole thing nonsense and said that the fault lay in the original formulation of the inverse square law.

Coulomb's formula for the force between two equal electric charges was based on the idea of action at a distance – that the charges somehow sensed one another out and produced a mutual attraction or repulsion that acted along the straight line between them. But this, in Heaviside's opinion, was eyewash. What really happened was that each charge infused the surrounding space with a radial field of electric lines of force,

[§] Also called the permittivity of free space.

or flux, and the field strength at any point was the flux density there divided by the constant ε. At a point a distance r from one of the charges the density of its flux was its charge, q, divided by the surface area of a sphere of radius r – that is, $q/4\pi r^2$. The force on a second charge of value q placed there would be q times the field strength from the first charge, so the proper formula for Coulomb's force between two equal charges was:

$$F = q^2/4\pi \varepsilon r^2$$

This was the natural home for the 4π factor. Put 4πs here and in the similar formula for the force between two magnetic poles and they vanish from everywhere else. The legion of 4πs that had seemed intrinsic to electromagnetic theory were, in truth, nothing more than correction factors made necessary by the omission of 4π from its proper place. A compelling case for change, one would have thought. But moving 4π would send ripples through the whole system of units; for a start, the new unit of electric charge would be smaller than the old one by a factor of $\sqrt{(4\pi)}$. Physicists and, for the most part, engineers had grown comfortable with their odd collection, and commercial interests were vested in the status quo. In short, there was massive resistance to change. This didn't stop Oliver making his pitch.[17]

> The 'brain-wasting perversity' of the British nation in submitting year after year to be ruled by such a heterogeneous and incongruous collection of units as the yard, foot, inch, mile, knot, pound, ounce, pint, quart, gallon, acre, pole, horse-power, etc., etc., has been repeatedly lamented by would-be reformers, who would introduce the common-sense decimal system; and amongst them have been prominent electricians who hoped to insert the thin end of the wedge by means of the decimal sub-division of the electrical units, and their connection with the metre and gramme, and thus lead to the abolition of the present British system of weights and measures with its absurd and useless arbitrary constants. But what a satire it is that they should have fallen into the very pit they were professedly avoiding! The perverse British nation – practically the British engineers – have surely a right to expect that the electricians will first set their own house in order.
>
> The ohm and the volt, etc,, are now legalised, so that, as I am informed, it is too late to alter them. This is a non seq., however, for the yard and the gallon are legalised; and if it is not too late to alter *them*, it cannot be too late to put the new fangled ohm and its companions right. It is *never* too late to mend.

The name Oliver gave to his proposed new system of units was 'rational'. Naturally, he called the standard set 'irrational'. When the idea first came to him he had been uncharacteristically tentative, using the new units when drafting his papers but converting back to standard units before publication. Then he introduced rational units in a short series of papers before reverting to standard ones. Finally, in the set of papers later published in book form as *Electromagnetic Theory*, he took the plunge, made his big pitch and never used irrational units again. After a while some prominent people added their voices to his: John Perry became a great enthusiast, and Sylvanus P. Thompson and W.E. Ayrton gave solid support. One by one, more people came over but the battle went on well into the twentieth century. Eventually Heaviside's cause prevailed. In 1960 his idea to relocate 4π was formally embodied in the SI (Système Internationale) set of units used by physicists and engineers all over the world.[18]

Everyone who studies or works in electrical science or engineering today should thank Heaviside for his part in banishing the troublesome 4π s. Yet, like so many of his achievements, this one has been lost from view. Even when his name *is* mentioned, it isn't always in a way he would have approved. One textbook author writes of Heaviside's relocation of 4π as a 'cunning device'.[19] An intended compliment, but one suspects that Oliver would have scorned it, saying that one might as well call the abolition of slavery a 'cunning device' – what he was doing was righting a wrong.

Sales of *Electromagnetic Theory* had not fulfilled expectations. When Volume I appeared in 1894, Oliver had breezily told Fitzgerald that he expected 1000 copies to be sold in the first year. In fact, barely 600 had left the shelves by the time Volume II appeared five years later, and there seemed little prospect of sales improving. Yet his ideas were spreading and his reputation as a sage was growing. It seemed as though many people were learning of Heaviside's work through trusted intermediaries rather than directly from his difficult papers and books. Fitzgerald and Lodge had played such a role and others such as John Perry, who had been glad of their mediation, were now taking the job on themselves.

Perhaps the most effective advocates of all were the editors of *The Electrician*. For years they had championed Heaviside, not only publishing his articles but promoting his ideas in their editorials. It had not been an easy post. First Biggs and then Trotter had had to stand up to interference from the editorial board and both had left after taking as much as they could bear. The same thing happened again when Trotter's successor, W.G. Bond, resigned in 1897 after only two years in post. He, too, had been a good friend to Heaviside and wrote to explain that he had seen it coming: he had failed to 'flatter and fawn' sufficiently and knew that his ejection was 'a mere matter of time'. The new editor, E.T. Carter, was soon put to the test when Oliver asked for a contract for a third volume of *Electromagnetic Theory* even though Volume II was still some way from completion. Carter very reasonably asked for a synopsis of the proposed book and got the following response.[20]

> Synopsis? Can't. The Lord will provide. He always does. Besides that, I know I can make a third volume and very likely a fourth as well. I can't say more synoptically than, that, broadly speaking, vol. 3 would relate to electric waves in general.

Then, after commenting politely but dismissively on some well-meaning suggestions from Carter, Oliver concluded:

> I am afraid you will think the above very unsatisfactory .... I can't help that, though I am sorry, being much indebted to the Editor of *The Electrician* in the past for the opportunities given me. The best I can do is to suggest that you give me *carte blanche* and I will try to make the best use of it.

Providence had sent him yet another editor willing to go out on a limb, or so it seemed. *Carte blanche* is exactly what Carter gave him but things still did not go smoothly. Articles for Volume III got under way in 1900 and *The Electrician* published around fifty of them in the first two years, but then Carter began to reduce the frequency at which they came out. This was not at all to Oliver's liking – at that rate he reckoned it would take twelve years to complete the volume. He protested but this time failed to get his way. The magazine stopped publishing the articles – his charmed life with

The Electrician was over. His next batch of papers was published by *Nature* and the material then went to the Electrician Publishing Company for assembly into the book, but they were inordinately slow in producing proofs and in the end Volume III took twelve years to produce anyway.

Carter was taken ill and died in 1903, aged only thirty-six. Assistant editor, F.C. Raphael, took over but served only three years in the post. His successor, W.R. Cooper, tried to make up the quarrel with Heaviside, asking him to contribute 'articles occasionally on certain subjects', but had a dusty reply. Oliver said he had never written 'occasional' articles and wasn't going to start now.

In 1904 a letter arrived from the Secretary of the Royal Society, Joseph Larmor. The Society had instituted a new medal to commemorate the inventor David Hughes, who had died three years earlier, and they wanted to present it to Oliver for his 'contributions to the mathematical theory of electricity'. They hoped he would be able to attend the ceremony in London in November. Here were echoes of the 'exhibition clause' he had taken exception to when awarded his F.R.S. Then he had accepted, on condition that he wouldn't have to go to London. This time he gave a straight refusal; even Larmor's offer to waive all the formalities, including travel to London, failed to move him.

Although Oliver set little store by honours, his door was open to congenial offers. Why turn this one down? At first he kept his reasons to himself but he let them out in a letter written many years later to the French engineer Joseph Bethenod.[21] He would have accepted the Society's top medal, the Copley, but thought that Larmor, whom he mistrusted, was trying to damn him with faint praise by offering a minor one. Also, although Heaviside admired Hughes as an inventor, even calling him 'a great man in his way', he had little respect for him *as a scientist* and thought the Society had done badly in choosing to commemorate Hughes when there were others far more worthy of the distinction. If they had named the medal after Fitzgerald he would certainly have taken it.

He did accept an honour the following year. It was an honorary doctorate from the University of Göttingen. As always, he found the formal side of things rather ludicrous but the grandiose citation in Latin amused him and he used it to conjure up mock-Latin expressions for entertainment, for example regaling Searle and his wife with the greeting 'Georgio Searlio et spouso. Salutem. Te igitur.'[22]

The Searles continued to visit him once or twice a year, and at Christmas in 1906 were alarmed to find him quite ill, shaky and yellow with jaundice.[23] Life at Bradley View had not gone well for him. The 'hot and cold' disease was now only one of many ailments that came and went, with no professional diagnosis or treatment, and his relations with the locals had gone from bad to worse. He felt, probably with some justification, that his brother Charles and wife Sarah, for whom gossip was a natural part of daily life, had helped to fuel the abuse that came his way. A notebook entry describes some of the 'mischief' and continues:

> That amiable fool C. is responsible for a lot of it. Telling people I am 'afraid to go through N.A.' That is his own invention, for one thing. Then he had no right to say it at all, or anything, without consulting me. Then it gets repeated and spread all over and I am insulted more than ever …. I mentioned to him that I had refused medal. No reason given.

He said contemptuously, he thought I was very foolish. Then he goes and tells it at home, no doubt with further expressions of contumely. Then his wife goes and talks to the Man. And then the hatchet faced man is impudent to me. 'He, He, He … He, He, He. Arnt yer going to take yer Medal? He, He, He!' Then I try to make complaint to C., but he is up in arms at the very notion of my having anything to say against the man …. 'Well, *I* think it's very amusing!' was his nice remark on my telling him of some beastly behaviour of boys outside. And then, no doubt, he went and made game of it at home.

He didn't mind too much when boys stole his apples but it was a different matter when they broke his windows and stopped up his sewage outlet pipe. Unkindest of all, in a jibe that must, however unintentionally, have been instigated by Charles, they would walk up and down the road outside the house, shouting 'Poop. Poop. Pupin.'

Charles' thoughtless indiscretions may have caused Oliver pain but in other ways he did his best for his difficult brother, drawing diagrams for Oliver's papers and, as we have seen, providing him with a mechanical piano. It was not easy to help Oliver. Sometimes he would take an ordinary act of kindness as an insult. For example, when Charles' wife Sarah and her sister went to have tea with him they took a loaf of bread. Oliver refused to use the loaf and it sat on his kitchen table until Charles visited several months later and threw it out of the window. Oliver's health was a constant worry for Charles and Sarah. The winter was a bad time for him and each was worse than the last. Something had to be done.

Sarah's sister Mary, who had never married, lived alone in a big house near the music shop in Torquay and could take in Oliver as a paying guest. They would both benefit: Oliver would no longer have to see to his own housekeeping and cooking, and the income would help with Mary's mortgage payments, which were seriously behind schedule. Charles made the arrangements and Oliver moved house. In July 1908, he wrote to Searle from a Torquay address:[24]

My continued illness has obliged me to move, to enable me to go through next winter. Garden my work just now. Clipping bushes and sawing up wood etc. $1\frac{1}{4}$ acre. Lots of trees, bushes, and flowers. Now at most luxurious stage.

The house in Torquay was called Homefield and it became Oliver's home for the rest of his life.

Chapter 11
A Torquay marriage
Torquay 1908–24

Homefield was in Lower Warberry Road, Torquay, a short but sharply uphill walk from Charles' music shop in Torwood Street. It was built on the side of the hill and had been designed to take advantage of the topography. You entered at what looked like the back of a modest two-storey house but there were stairs leading down as well as up and the front of the house was much more imposing, with three storeys, ornately decorated gables and big bay windows that looked southwards over a long, down-sloping garden.

Mary had, we assume, inherited the house from her family but had also acquired a mortgage that demanded bigger payments than she could afford. She had once been better off but speculative investments had gone wrong and she had been the victim of her own good nature, spending generously and lending money to people who couldn't or wouldn't pay it back. The mortgagee had threatened to foreclose the loan on the house and Oliver's financial contribution was very welcome. But he, too, had been living beyond his means. He was nine months behind with the rent when he left Bradley View and brother Arthur had had to step in to make it up.

So Mary and Oliver, both of them hopeless at managing money, or anything else for that matter, set up house together, hoping that two could live as cheaply as one. Oliver called it his 'Torquay marriage' but romance didn't come into it. The first thing to settle was territory. Oliver took the top floor and Mary had her bedroom on the middle floor, next to the drawing room and dining room. The kitchen and scullery were on the lower ground floor which opened on to the garden. As the paying guest, Oliver regarded Mary as his housekeeper as well as his landlady, and took it upon himself to keep her 'up to the mark', as he put it. When doing his own housework at Bradley View he had happily adopted the 'principle of least action' but things were different now he had an assigned housekeeper: it was his duty to make sure that Mary did hers.

Oliver claimed to have Mary's own interests at heart and, in a sense, he did. She was a few years his senior but he seemed to regard her as a sweet but spoiled child who had never grown up. He referred to her as 'the Baby' and appointed himself as her guardian, protecting her from the 'flatterers' who sponged off her and seeing that she wore sensible clothes in cold weather. He once asked Mrs Searle to check that Mary was wearing warm enough underclothing.

Good food and home comforts worked wonders for Oliver's health but the years of hardship at Newton Abbot had taken their toll and he was always on the lookout for early symptoms of his various complaints so that he could try to stave them off. One of his preventive measures was to wrap up even warmer than usual and turn his gas fires even higher. His love of heat was a source of domestic discord, especially as Mary had been in the habit of opening outside doors, even in the winter, and forgetting to close them. They also fell out over what to have for dinner. Brawn was one of Oliver's favourites; pease pudding was another. The pease pudding and lentils incident has become almost as well known as Heaviside's battles with Preece. He described it in a letter to Searle.[1]

> The great lentil Question cropped up today (not for the first time). Shall I want Pork and Pease pudding hot, this being the proper time for that wholesome and vulgar fare, to make the system able to resist the cold, shall I be diddled into eating lentils instead on the plea that they are much nicer, and *so* nutritious? Never! I had enough of it before. I was introduced to lentils at Paignton by a niece who took charge when my mother became too feeble; it was substituted for my mother's pease pudding; most unwarrantably and without any consideration for our feelings or wishes, but merely because this new cook was a vegetarian, and vegetarians seem to have a spite against pease and always preach lentils. Why? I hardly know, probably because they have been proved by chemical analysis to contain a little more nitrogen than pease. This learned girl (a woman now) had *nuts* for breakfast, because they were recommended by some idiotic vegetarian! It's a mad world. I preferred the pease but never had 'em again. It was always that sloppy lentil soup.
>
> But why does the Baby do it? She isn't a vegetarian, eating nuts for breakfast, with vegetarian butter (a fraud), and vegetarian cheese (another fraud) at other meals, all very nutritious and nitrogenous, no doubt. Because she was once strongly under the vegetarian niece's influence, and so imbibed a lot of her nonsense, and it hasn't gone off yet. I have, however, got rid of cabbage and stalk soup, and some other wretched frauds. She eats real good cheese now Cheddar and St Ivel, and all sorts of non-vegetarian food. (Perhaps too much).
>
> Having asked for the seasonable dish (a change from chopped up steak and potatoes – half black), I got the pork because there was some in the house, rather stale, and not the right sort, but wouldn't have the lentils or their nutritiousness. (Several times same thing before). She wasn't amenable to my very civil remonstrances that I knew lentils very well; I wanted pease. 'Oh! You know everything!' She is going to buy some, if procurable. To keep her from forgetting I drop down a note periodically. No.1 (new series) informed her that the Jews ate lentils in the Bible, but there is no mention of pease pudding. No.2 (in preparation) there was a plague of lentils in Egypt at the time of Moses. Also there was one case of living for forty days on lentils and wild honey, or else honey and wild lentils, they were so nutritious. No.3 (ready tomorrow) mentioned in Magna Carta, Felony to rob the villein of his pease pudding. No.4 (soon) Act of George IV. Fine 40/- or one month on grocers and others for substituting lentils for pease pudding. And so on. I shall get my pease pudding in time, as I did my Brawn. That's another story.

Oliver's letters to Searle are full of accounts like this and would furnish an almost complete script for a TV sitcom. One can picture the characters – the plump rosy-cheeked West Country landlady and her cantankerous, mad scientist lodger from London – having their rows and reconciliations. But reality was somewhat darker. Oliver was becoming a tyrant. Obsessed with the notion that Mary was neglecting her duty, he tried to stop her friends from visiting and even objected when Mrs Searle took her out to a concert. Searle told him in the plainest terms that he was behaving abominably, repaying Mary's kindness with cruelty. Oliver responded with silence and we can only guess his thoughts and motives. Perhaps the nearest one can get to an explanation is that he simply lacked the faculty of being able to rub along with people. Relations, friends and colleagues had to take him on *his* terms, even if those terms sometimes appeared harsh, even cruel. There was no malice, just an idiosyncratic, and occasionally perverse, sense of what was right.

Whatever the reasons for Oliver's behaviour, it didn't stop his friendship with the Searles. On some visits Mrs Searle's sister Emma Edwards came along. She and Oliver got on famously and he often asked after her in letters to Searle, abbreviating her name to E^2. Mrs Searle's other sister, Amy Edwards, also came to Homefield. In fact she visited once on her own and reported that she had enjoyed herself very much. By the strict conventions of the time it would have been improper for Oliver to invite her, so the invitation had come from Mary and they had entertained her together – a demonstration that they didn't fight all the time.

Oliver's creative powers had not escaped the ravages of illness and he found it impossible to concentrate on work except in short bursts. In conversation or in letters he was still sharp as a knife but the kind of sustained intense thought by which he had made his discoveries was now beyond him. He told Searle 'my mental activity is gone for good' but this didn't stop his extensive correspondence with men of science all over the world and his reputation as a great sage continued to grow. Homefield, home to a curiously matched couple, quarrelling and eking out their meagre resources, was leading a parallel life. It took over from Bradley View as the temple of wisdom and expanded its operations, even acquiring its own brand-name, the 'Inexhaustible Cavity'. It is not clear how the name came up but it caught on. According to one story, a letter addressed to 'Inexh. Cavy. Torquay' arrived safely.[2]

In one historically important sense Heaviside had, himself, arrived. The Institution of Electrical Engineers elected him an honorary member. This was tantamount to a formal recognition that mathematical theory was part and parcel of electrical engineering. It had been slow work, but he had transformed the profession. The days of the 'practical' men who derided theory were over. In its original form as the Society of Telegraph Engineers, the Institution had at first turned him away because it didn't want 'mere clerks' as members. A little later it had thrown him out for not paying his subscriptions. Now, thirty years on, it offered an olive branch. Perhaps Oliver recognised the historical significance of the official reconciliation; at any rate he accepted the honour without, for once, raising objections or placing conditions.

In 1912 he was short-listed for the highest award of all, the Nobel Prize for physics. He would certainly have accepted this one because it came with a pot of cash big enough to sweep away all his money worries, but he was not selected. He was in good company: among the other unsuccessful nominees that year were Albert Einstein, Hendrik Lorentz, Ernst Mach and Max Planck. The Nobel Committee gave the Prize to Nils Gustaf Dalen, who had invented a way of feeding gas fuel automatically to lamps in lighthouses and buoys. Their choice may seem curious to us but it shows the immense importance then attached to saving lives at sea.

The long-delayed Volume III of *Electromagnetic Theory* came out in 1912. A sure sign that Oliver was weary of the business of publication was the fact that he wrote a preface of only 120 words – for Volume I he had written 3000 words. But the pen was as sharp as ever. He wrote:

> Long ago I had the intention, if circumstances were favourable, of finishing the third volume of this work about 1904 and the fourth about 1910. But circumstances have not been favourable. That is all that need be said about it here, save to add that I have excluded parts of the third volume and included parts of the fourth.
>
> It would be as wrong to love your enemies as to hate your friends. Nevertheless, 'the way of life is wonderful; it is by renunciation.' … If my life is spared, I hope to be able to present a bust of the eminent electrician who invented everything worth mentioning to the Institution over which he once ruled, to be placed under that of Faraday.

Fittingly perhaps, Oliver's last published comment on William Preece was a mock-gesture of reconciliation. As a young man, Preece had helped Michael Faraday with some experiments. A master at self-promotion, Preece never let a good opportunity go to waste and he exploited this one to the full, telling everybody how he had 'sat at the feet of Faraday'. The Institution of Electrical Engineers had a marble bust of Faraday in its grand new headquarters in Savoy Place. The idea of placing another bust underneath it, where it would be knee-high to people walking past, was, of course, ludicrous. British readers would have seen the elaborate in-joke at once but goodness knows what everyone else made of it.[3]

William Preece died the following year. He was acclaimed as a great man and received about 200 obituaries in the press. Many testified to his benevolence and kindliness, no doubt on good grounds even though these qualities had not shone in Heaviside's direction. Oliver had sport annotating the newspapers' comments with his own. To the statement that any subordinate who had a technical improvement to suggest would find in Preece a sympathetic listener, Oliver added 'to steal from if he could'. Perhaps the pick of the comments was in the *Daily Telegraph*. Their writer said:

> He inspired so much confidence in the official world and among his colleagues that it became a motto, 'Preece is always right'.

We don't know Oliver's response to this one, but it must have reinforced his long-held agreement with Thomas Carlyle's best-known dictum. When asked the size of the British population, Carlyle had replied 'thirty million, mostly fools'.

Like Volumes I and II, Volume III of *Electromagnetic Theory* was mostly an exposition and development of ideas that Heaviside had introduced in his earlier papers. But they were profound ideas and the three volumes were still brimful of originality. One theme he continued to pursue at length was the shape of the field surrounding an electrical charge. As we have seen, he had worked out that the field would contract along the direction of motion as the charge approached the speed of light. Foreshadowing Einstein's special theory of relativity*, the mathematics suggested that nothing would be able to exceed the speed of light, but this didn't stop Oliver investigating what would happen if the charge did. He predicted that it would emit radiation in the form of electromagnetic shock waves with a cone-shaped wave front, something like the sound shock waves we hear when an aeroplane breaks the sound barrier.

---

* Heaviside's first results on this theme came out long before Einstein published his special theory of relativity in 1905.

By the 1930s it had became clear from Einstein's theory and supporting experiments that nature did indeed impose a speed limit on all material bodies, and that this limit was the speed at which light travelled in a vacuum. Heaviside's speculation was beginning to look like science fiction. But, amazingly, it turned out to be fact. In 1934, the Russian physicist Pavel Cherenkov found conical electromagnetic shock waves, exactly as Oliver had predicted, while studying the movement of charged particles in water. The conical wave front formed because Cherenkov's particles were moving faster than the surrounding water could transmit the waves. This was possible because light travels more slowly in water than in a vacuum whereas nature's speed limit on any material body is the same no matter what medium it is moving through. Cherenkov and his colleagues Ilya Frank and Igor Tamm were awarded the Nobel Prize for physics in 1958 but they made no mention of Heaviside, who had published his first study of faster-than-light particles in 1888.

This was yet another instance of Heaviside's pioneering work going unrecognised by posterity. But he hated false aggrandisement and it would be doing him no service to hail him as an all-seeing sage, everywhere ahead of his time. In one respect he was by now falling behind, anchored by the very concept that had served him so well – the field. Like Faraday and Maxwell, he believed that electric and magnetic energy were borne by stresses that existed in *non-conductors* (including empty space). In his scheme conductors played only a secondary role, as materials that could not support the stresses of the field and hence held no energy. A conducting circuit acted as a channel along which the stress was released. Charges and currents were simply effects produced in conductors by stresses in the surrounding field, and the same stresses gave rise to electric and magnetic forces. All this was in complete contrast to the earlier view that electricity was something that existed only in conductors and that electric and magnetic forces resulted from the charges or currents in the conductors acting on one another at a distance.

The field view had triumphed by predicting electromagnetic waves and explaining other phenomena, like the skin effect. But in 1897 Maxwell's successor but one at the Cavendish, J.J. Thomson, discovered the electron – a material particle with an intrinsic electric charge. A little later the American physicist Robert Millikan confirmed Thomson's results and measured its electric charge and mass. Nature had sprung a surprise: the electron was its unit of electricity and it was *a constituent of matter*. It seemed that the field theorists were wrong after all – an electric current was a physical flow of electrons along a conductor and a charged body had either a surplus or a shortage of electrons on its surface.

Heaviside never came to terms with the electron. Despite the experimental evidence he couldn't bring himself to think of it as anything more than a hypothesis. Maxwell's electromagnetic theory was entirely a field theory, with no place for electrons, and Heaviside was not prepared to modify the theory to let them in. Others were, notably Hendrik Anton Lorentz in Leiden and Joseph Larmor in Cambridge, and in the end the electron slotted in neatly.[4] Oliver's main stumbling block was a chicken and egg question: does the electric charge cause the field or does the field cause the charge? Maxwell and his followers thought the field came first, but the electron seemed to require things to be the other way round. For Heaviside, primacy of the field was a principal tenet of faith and he couldn't give it up.

It turned out to be a false dichotomy. Physicists came to the view that both the electron and electromagnetic fields were fundamental – there was really no point in assigning cause and effect. With this interpretation, they adapted and extended Maxwell's theory to accommodate the electron. In doing so, they went beyond Heaviside's territory; he had opened up great tracts of electrical theory but his pioneering days were over.

In the preface to Volume III of *Electromagnetic Theory*, Oliver had tantalised his followers by making an enigmatic reference to Volume IV. It seemed to imply that he had given up the thought of producing a fourth volume but there was enough ambiguity to keep the door open a crack. He did carry on writing and at one point had an offer from an American enthusiast, Dr Ludwik Silberstein, to come to Homefield and help him prepare Volume IV for publication. Oliver turned down the offer, saying that all he had was raw material, resembling a scrapheap. There may have been gems buried in the dust but it was a fair description of the state of his last, unpublished, papers. There was to be no fourth volume.[5]

In 1911 Oliver had become the owner of Homefield. The business was arranged by his brother Charles' wife Sarah, who was Mary's sister. One object, clearly, was to relieve Mary of the worry of keeping up the large and high-maintenance property but another was probably to straighten out Mary's financial affairs, which had for years been managed by Kitson's, her solicitors. On the face of it the outlook was bleak. She owned several other properties that were rented out but all were mortgaged and at least one of the tenants was well behind with his rent. She had also been given a further large loan by Kitson's. Yet the only interest she paid directly was that on the Homefield mortgage. Oliver suspected Kitson's of sharp practice at the expense of their over-trusting client. He commented in a note found among his papers: 'They pay themselves by unknown means. Very dark.'[6]

In his own way, Oliver wanted to help Mary, too, and he borrowed heavily to buy the house.[7] Oliver was in debt for the rest of his life but he did what he could to pay his creditors and regarded any new debts as personal liabilities rather than extra charges on the property, which was to be left to his nephews and nieces, even though they might need to sell it to clear his remaining debts.

Perhaps the stress of completing the business with Kitson's had sapped Sarah's strength; soon afterwards her health began to fail and she became very ill. Mary was distressed and became more upset every time she visited her sister. Oliver worried about Mary's mental state (pots and kettles!) but had a bizarre way of showing his concern. Convinced he was protecting her from further anguish, he put a stop to Mary's visits to her sister and when Sarah died he even refused to let Mary go to the funeral.

The following winter Oliver was himself desperately ill. He was never completely free of ailments while at Homefield but this attack was the worst since his last winter in Newton Abbot and he was laid up for several months. Charles hired a nurse for him but he took against her and for ten weeks she did little more than peep round the door to his room several times a day. The nursing was left to Mary. At one point the nurse said he was going to die but Mary knew better and replied 'He won't die; he will live to turn you out of the house.' She was right. By May he was fit enough to get around the house and the nurse had gone.

Oliver and Mary continued their curious partnership. Here Oliver describes a typical episode to Searle. In explaining some of Mary's shortcomings he reveals, perhaps, more about his own.[8]

> What cast some humour on the situation, of a grim kind to an invalid, was my finding the doors of the two rooms shut which I had particularly requested the Baby to keep open, all the year round! There is no draught, and I want the warm air from the gas stove to enter them dry, or as little damp as possible. I like to have a good grumble occasionally. I would not do so if the evil were natural or essential. But it is easily preventable. It used to be far worse the first year I was here; if I put a lump of coal on the fire, the Baby took it off again (not always), and told me to go away to my room. But how glad she was when I gave her one of my gas stoves to warm her bedroom, when she was suffering from a most severe chill internally, besides rheumatism, gout, eczema, swollen joints. The relief was so great, and her improvement so rapid (internal chill and consequences) that she was really grateful to me. For a time. All forgotten now. As for my being ill then, she had hardly any appreciation of it. Same with others. My ulcerated bleeding stomach, and jaundice, etc. became 'a little poorly', or 'a little tightness on the chest', or 'HAVE you been Out Today?' again and again. Especially the formula 'That's because you this that or the other', at fancy. Or else 'Ah, you don't etc., etc.' Or it may be the hard stare, and then 'I don't see anything the matter with him. I don't believe there is. All imagination.'
>
> But I am wandering from the mark into a region requiring volumes to describe. The great lesson of my life is that it is MONEY, and nothing else that rules the world of Common people, that is, nearly everybody.

The great lesson of his life? Hardly. More a cynical observation while in a dark mood. Nevertheless, money worries dominated daily life. Oliver was sometimes forced to borrow from Searle: one letter shows that he owed his friend £60 and was asking for another £20, at the same time making it clear that he would not take a gift. Some desperately needed relief came in 1914 when his Civil List pension was increased from £120 to £220 a year but, even then, the household's financial position remained precarious.

One way to economise, Oliver thought, was to curb Mary's excessive generosity to her friends. He felt sure that some of them saw her as a soft touch, and took it as his duty to keep spongers away. One friend, in particular, used to stay overnight, 'making a boarding house of the place', until he got rid of the spare bed. Mary was not pleased to have this kind of protection and they had many arguments on the subject. After one particularly bad row Mary downed tools and kept to her own part of the house, leaving Oliver to do his own housekeeping. Just at this time Oliver had an unexpected visit from one of Searle's colleagues at the Cavendish, Arnold Crowther. The visitor later recalled his experience to Searle, leaving him, and us, with an illuminating snapshot.[9]

> I seem to remember a rather thick-set, rosy person with a good deal of hair about his face, and dressed rather shabbily and untidily, even by my standard which (having been accustomed to distinguished dons) was not a very high one.
>
> The room I was shown into was large and very brown. There was a large table in the middle, littered with books, manuscripts, and the remains of a solitary lunch, all mixed up, and on the sofa a similar mélange of books and MSS together with the remains of the breakfast crockery. I have an idea that relics of still earlier meals were somewhere in the recesses of the room, but I cannot be sure. As far as I could tell, Oliver was living in

the big old-fashioned house quite alone at the time; at any rate there was nothing to hint of any other occupant, or any sort of attendant.

Heaviside immediately launched into a long argument on some electromagnetic topic, from which I gathered that all was not well with the commonly accepted treatment of that subject – an impression which I retain to the present day. From time to time I interjected what I hope were appropriate remarks – but in spite of your careful training I was, and still am a little shaky on mathematical electricity. After half an hour or so I was kindly but firmly dismissed; and I have never had the courage to call again.

I was, I may add, very deeply impressed by Oliver Heaviside, even in such a brief encounter. I have never met anyone who (in spite of surface eccentricities) impressed me more deeply with the feeling that I was, momentarily, in contact with a really great mind. I have always been glad I made the visit.

Mary and Oliver patched up their quarrel soon after Crowther's visit but the combative life was wearing her out. Suffering from persistent gout as well as battle fatigue, she eventually 'sank into a state in which she sat and stared into the fire', as Searle put it. She had had enough. One day her nieces called with a car and took her away to their father Charles' flat above his music shop in Torwood Street, where she soon recovered her spirits and composure in the calmer surroundings. Mary's nieces were Oliver's too, of course, and no doubt felt a twinge of sympathy for their Uncle Olly even as they rescued Mary from his oppressive regime. One of his young relations probably summed up their views well when she affectionately described him as 'an awkward old cuss but a brilliant man'.[10] Not that they knew what he was brilliant at, except that it had something to do with electricity and mathematics but, like the recent visitor, Dr Crowther, they could see the great man beneath the 'surface eccentricities'. They also knew how difficult he was to help and, for the most part, stayed away from the house

Something that became apparent to everyone over the next few years was that his 'surface eccentricities' were deepening. Before long, even the kindest observer would have described him as potty.

He had a running argument with the Torquay Gas Company, whom he called the 'Gas Barbarians'. He was usually late paying his bill and they cut him off several times. He used enough gas for twenty normal households – one year his annual bill came close to £200 – but always wanted more and blamed the company for the feeble supply. To try to get more heat into the house he used to improvise with the pipes, adding flexible tubes which he repaired with putty, paper and string. One day he thought the supply was becoming choked by debris in the main feed pipe, so he set out to clear it by unscrewing a blanking cap and letting the gas stream through. To prevent an explosion he lit the jet, getting a flame several feet long which he then put out with a wet cloth before replacing the cap. He was probably lucky to escape serious injury but the escapade left him with painful burns on the face and hands. The Searles happened to call a few days later and he came to the door dressed like a Tuareg from the Sahara, peering with one eye through a chink in the blanket that covered the rest of his head. For all his physical afflictions, he had a tough constitution, and the burns healed quickly.

It seemed to him that most of the world was unappreciative of his achievements and insensitive to his needs. But this was to be expected given that the population were 'mostly fools' and it could be borne with a shrug of the shoulders. He had kept his independence intact and that was what mattered. And as most of the world seemed to regard him as a crazy misfit he might as well play the part to the full. In mock-solemnity he awarded himself the enigmatic title W.O.R.M. and put it after his signature in letters, not only to Searle, who played along, but to strangers, who were mystified. He loved to receive letters addressed to Oliver Heaviside Esq., W.O.R.M. from people who had taken it to represent some distinction. And to take the joke further he sometimes signed letters with a near-anagram of his name: O He Is A Very Devil.

While he was having his fun with mock-honours the world decided to add to his real ones. Early in 1918 he had a letter from the Secretary of the American Institute of Electrical Engineers, B.A. Behrend, asking him to accept honorary membership.[11] It was indeed an award of distinction – he would be only the fifth man to be so honoured since the Society was founded nearly forty years before – but if Behrend expected a reply on the conventional lines of 'Thank you. Yes.' he was in for a shock. The citation praised Oliver's contribution to the huge improvement in telephony brought about by inductive loading – putting high inductance coils at intervals in the line. But the generous-seeming words struck a raw nerve. The taunts of 'Poop. Poop. Pupin.' from the street boys in Newton Abbot rang again in Oliver's ears and he wrote back:[12]

> I took notice that although no one questioned that the loading coil system was my invention *before* Pupin sold his patent (though it was sometimes laughed at by practicians who miscalled themselves practical men), yet *after* a great change quickly occurred and I was repudiated in a most emphatic manner. Now, being an innocent in relation to dollars, from want of experience possibly, it may be malignity on my part to suspect that the reason of the change was based on commercial considerations. Prior publication, of a distinct kind, if admitted, might have interfered with the flow of dollars in the proper direction … What is the practical remedy? To perform, in my idea, a belated act of bare justice. If your great Institute, through its head or friends, is prepared to announce … that I am the inventor of the loading coil system … I don't care to be tacked on to Pupin, either fore or aft. Pupin got his dollars, and I do not see what more he is entitled to, save relatively insignificant 'improvements' in practice. I want explicit recognition, as well as implied.

Oliver must have known that it didn't matter how strong his case was – there was no chance of the Institute agreeing to a public humiliation of one of its own most distinguished members. Presumably, he felt compelled to pursue the point on principle. After exchanging several letters with the patient but immovable Behrend, he accepted the honorary membership on the terms first proposed.

Oliver wrote and received many letters and one of his correspondents, Dr C.V. Burton, an Oxford physicist, became a good friend. Burton visited Homefield when Oliver was recovering from one of his severe bouts of illness and was shocked to find him acting as his own doctor. Like many others had done, Burton tried, and failed, to persuade Oliver to seek proper medical advice. When the First World War came, Burton went to work at the new Royal Aircraft Establishment at Farnborough in Hampshire. He and Oliver discussed the war in their letters and took opposing views. Burton saw it as 'a calamity' but Oliver, in his perverse way, argued that it had had a bracing effect on the country. His letters to Burton have been lost but we can get an idea of his attitude from a postscript he added to one of his letters to Behrend, in the knowledge that letters to America were liable to be opened and censored.[13]

For information of Censor. Kaiser Bill is one of the best friends we have ever had, because
he is waking us up.

How did he come to this extraordinary view? We can only assume that he had seen
the people of Torquay shaken from their comfortable preoccupation with local events
and gossip, and thought the change beneficial. At that time neither he nor anyone
else outside the Government and the army high command knew the full extent of
the slaughter in the trenches. Oliver lost his own friend in the war. While working at
Farnborough Burton was killed in an accident with the poisonous gas phosgene.

Even with the increased pension, Oliver never managed to haul himself out of
his financial slough: on top of personal loans his standing overdraft at the bank often
exceeded £100. He was in constant fear of being 'sold up', and the threat was real.
One day a police inspector came and put tickets on his furniture, which was to be
taken away and sold the following day. Only at the last minute did Oliver manage to
persuade the bank manager to extend his overdraft so he could pay the creditors.

As with other aspects of his life, he was partly the cause of his own troubles.
Admirers from all over the world wanted to help and had he been able to accept gifts
gracefully he could have lived his last years in comfort. But the thought of being cast
in the role of beggar appalled him. Back in 1894 he had refused an offer of money
from the Royal Society's fund 'for the aid of such Scientific men or their Families as
from time to time require and deserve assistance', and the pattern was repeated again
and again.

Some money did get through. There was Perry's £100 in 1896 and later several substantial sums came from France, organised by the leading engineer Joseph
Bethenod. But Oliver had acquired a formidable reputation for being difficult to help
and most people feared to approach him directly. Searle did his best to act as broker.
People would send him cheques, which he would pay into Oliver's bank account
while trying to concoct explanations that would satisfy his sceptical friend. Searle
performed the task nobly, getting no thanks from Oliver, who berated him for writing
'begging letters' on his behalf.

Searle heard in December 1920 from Frank Gill, of the International Western
Electric Company, that several of the big American engineering companies would be
willing to help Oliver Heaviside if they could.[14] Here was an offer that would banish
Oliver's money worries if only he could be persuaded to accept. Unfortunately, Searle
muffed his chance. Many years later, he painfully recalled their conversation.[15]

On the morning of the first Sunday after Christmas, I went to Homefield and said that I
wanted to talk seriously to him. He said 'What about?' I said I had a letter from a man.
'What man?' he said. I said I did not feel authorised to give the name. I said the man thought
he could get some money from companies in America, and that he wanted to know if Oliver
would be willing to accept it. 'I demand full particulars' he said. He was quite unwilling
to consider anything of the sort. 'I must make some reply', I said, and he constructed the
rudest thing he could construct. I said I could not send that, and we agreed on something
milder, and this I sent to Mr Gill.

Another scheme to help Heaviside had foundered. Searle came to realise that he
had handled things ineptly. 'I had a letter from a man' was the worst possible way
to begin. He should have thrown discretion to the winds and come straight out with

something like 'Frank Gill tells me that companies X, Y and Z would like to give you some money in recognition of the extra business you have brought them.' But the damage was done, and it was not just to Heaviside's finances. Oliver felt badly let down by the way Searle had refused to take him into his confidence, and when the Searles called to see him the following Christmas they found a note outside the door saying he wouldn't be able to see them that day. They called every day for two weeks but were turned away each time. And the same thing happened the following Christmas. As Searle put it, his discretion cost him dear. It was not the usual kind of falling out between friends. Oliver's rejection notes were written in his usual perky style with no sign of sulk. He just felt that, until Searle told him who 'the man' was, the friendship would have to be put on hold.

Oliver had by now acquired a local friend from a surprising quarter. One of the duties of the local policeman, Constable Henry Brock, was to serve warrants for payment of debt and this had brought him many times to Oliver's door. What had started with constabulary duty became something rather different. Bobby, as Oliver called him, gradually took on the role of Heaviside's minder, calling in most days, running errands, and generally seeing to his welfare. Brock's technique in conversation with Oliver was to make sure that his meaning was received and understood – 'he roars at me louder and louder, explains things to my stupidity, etc.' – but he clearly held affection for the old gentleman and the two got on handsomely. Oliver had a weekly delivery of groceries from Lipton's shop in Torquay but Brock used to buy or borrow many other things that he needed, including, on one occasion, a galvanometer. One of the Torquay tradesmen told Searle that Brock often paid for these items by tapping a mysterious local benefactor. The Brock family seem to have treated Oliver like an aged relation; he wrote them letters and Brock's daughters, Bessie and Kate, used to call in occasionally to help straighten out the house.

Oliver's two personae were curiously matched – the odd resident of Homefield and the internationally renowned scientist Dr Oliver Heaviside, F.R.S. And while one grew odder the other grew grander. In 1922 the President of the Institution of Electrical Engineers, John S. Highfield, wrote to say that they would like him to be the first recipient of their new, and highest, award, the Faraday Medal. Oliver's response to such offers had become almost routine: he didn't set much store by honours; there were too many of them nowadays; if he accepted it would be on condition that he was explicitly recognised as the inventor of inductive loading in telephone lines. In the end he did accept and, as the Institution knew that Oliver wouldn't come to London, they proposed to send 'a deputation' to Torquay to present the medal. A deputation! No chance. Oliver replied:[16]

> Alone, or with a lady to protect you against my notorious violence …. I usually get on well with ladies, with clear soprano voices that are so distinct and so unlike the throaty voices of gruff men. And they like me too, I think, even too much, though I don't flatter them. But they must have some sense of humour …. No deputation. A lady for protection allowed.

No grand ceremony in London and no deputation. But it was an important occasion and the Institution wanted to conduct things in a proper style. Highfield made two preparatory visits and a third to present the medal. For the first visit, he sailed to

Torquay in his yacht and cautiously sent his skipper to make the first contact at Homefield. An hour later his man was back at the harbour, having failed to get any response to his knocking on the door. The next attempt was more successful and Oliver explained that he didn't always hear knocks – the skipper should have blown his whistle through the letterbox like Bobby did. Highfield did his duty and in reporting the results gave us a picture of Oliver in his last years.[17]

> Heaviside lived entirely alone in a pleasant house in Torquay – a house decaying from long neglect. I called first by appointment and found him waiting in the weed-covered drive in an old dressing gown, armed with a broom, trying rather vainly to sweep up the fallen leaves. He was pleased to see me in a queer, shy way and took me through a furniture-laden hall, all covered with dust, to his own room. He had papered the walls with prints and reproductions from many publications, and pointed out to me old Presidents of the Institution of Electrical Engineers and all the recent ones, and asked me all about them. The walls held a pictorial record of his life's interests. He was in all ways entirely competent and preserved his power of impish criticism. Often he was amusing in a caustic way; he was genuinely pleased that he had been recognised as one of our famous men. His way of life made a pathetic background to his mental activity, but I am sure he did not regard it as pathetic. He talked freely of all recent work and development, but his talk was interfered with by many personal complaints of the most homely description. I saw him several times and tried to improve his ménage and especially his food supply, which was inadequate, but with little result. He was quite content to live as an old-world hermit, and so indeed he lived.
>
> He vigorously criticised the wasteful expense... but was consoled by the Medal being of bronze and not of gold. He read every word of both document and Medal and was especially pleased to see the name of Alexander Siemens, who signed as the oldest Past President. He talked much about telephony and wireless, all interspersed with homely grumbles at the many defects of his neighbours. He seemed to know all that went on in the town. It is impossible to give any adequate account of one who so despised what most men desire, but when I left him I felt that he was content, that he respected the Institution, and that it had pleased and made happy one of its famous men.

Highfield and his colleagues at the IEE had backed up their concern for Heaviside's welfare with action. The Chief Clerk, R.H. Tree, travelled to Torquay several times, on one occasion bringing 'a fine fat sole', which Oliver cooked 'on a special fire of Lipton's thin boxes and found delicious'. The two men took a liking to each other and Oliver had another friend. There was financial help, too. When Oliver was having one of his gas bill crises Tree found a way to get him to accept a cheque for £100 from the Institution. The letter, signed by Highfield, said:[18]

> The Council desires me to ask that you will be so good as to accept the action which they have taken as a slight expression of their high appreciation of the valuable services rendered by you to Electrical Science and Engineering, and to add that they leave it entirely to you to decide whether you will accept this sum, or whether you would prefer to regard it as a personal loan.

It had taken many attempts and many years, but at last someone had found a reliable formula for persuading him to accept a gift with good grace. It was simply a matter of recognising that the only way Oliver Heaviside would willingly conduct business, whether giving or receiving, was on his own terms. Had others found the formula earlier – from the Royal Society in 1894 to the American electrical engineering companies in 1920 – they might have succeeded in their wish to ease Oliver's path through life.

The path was now near its end. Oliver's brother Arthur died in 1923 and Charles in 1924. He was soon to follow, but not before he had gained another friend and got into another fight.

American universities were streets ahead of their British counterparts in bringing Heaviside's methods into their courses. At Cornell, Professor Vladimir Karapetoff invented a machine called the Heavisidion for solving the equations of transmission lines. When he heard of it, Oliver expressed surprise but said that it didn't provoke him to commit Heavicide. One of the most vibrant departments of electrical engineering was in Union College in New York State. The department's founder, the brilliant, hunch-backed Charles Proteus Steinmetz, had been a great admirer of Heaviside, and so was the current head, Professor Ernst J. Berg.

In 1924, Berg made the long journey to Torquay to visit the great man. Oliver was pleased to hear the latest news from America – engineers were taking enthusiastically to the operational calculus and a new edition of *Electromagnetic Theory* was on sale at $35 a set – but to Berg's disappointment he wouldn't discuss his writings, saying he had forgotten all about them. Even so, they found plenty to talk about. Like all new visitors to Homefield, Berg was shocked at what he found. Oliver's physical condition was especially worrying. It was June and he had recovered from the winter but he still had a formidable array of ailments. His feet were in a dreadful state and he didn't have any suitable shoes, so when Berg arrived home he sent a pair of American boots that he thought might do the job. They must have cost more to send than to buy, and transportation by ship was slow, but they eventually arrived and Oliver was grateful. His reply, written on 16 November, was one of the last letters he wrote and shows him as ready as ever to fight for justice as he saw it.

> I duly received the boots marked '5 dollars' and I think they are as near as possible perfect without orthopaedic operations …. The arrival of winter has caused Gas Co. trouble again. Hardly ever enough gas and rapid failure of the old, small mains by naphthalene and other chokage.
>
> Worse than that I had to have some repairs done to the windows and roof slates and gutters, to keep out the wind and rain. The single man I got was most willing, but was too confident, and he was not well, though he said he was. He was quite grateful for my practical way of helping the out of works, by giving them work they *can* do, treating them liberally as well. But he didn't understand the construction of the roof (his master does mostly repairs to small workmen's houses). So I had to direct him in many respects, and be out of doors, and I had an accident. In coming down a ladder my coat caught in something, and down I went, a fall of 11 feet on the broad of my back. The man did the usual stupid thing by trying to get me on my feet to see if I had hurt myself. I said let me lie, till the pain in the back goes. In about five minutes it was gone, and I was able to be pulled up, and actually walked indoors shakily. The pain had gone through me to the front, to the muscles

and bones all the way down from the top-of-the-chest. Well, by warmth and in bed I would cure that in a few days. But I could not get the warmth or proper food, and the window mending had to go on, so the house chilled and the man had to be looked after. The result was a violent attack of internal derangement …

I have another fight on. I order Milkmaids Pure Cream with no Preservatives. Liptons send me the common heavily sugared and adulterated tins they call 'Milks'. I refuse them. They try to make me take them. I write to Switzerland. They sent my letters to London. The London firm said they were greatly obliged to me. They were astonished to learn that Liptons didn't stock the Pure Cream Milkmaid Brand and would do their best for me. Liptons made a special requisition for me and got 24 Milkmaids from their London firm. With lump sugar added according to taste they are splendid, and I ordered 48 more, with 6 pounds of lump. They sent the lump but no more Milkmaids. Yet I have to proceed cautiously. I have 167 'Milks' and I want them taken back in exchange for 'Milkmaids' and I won't pay their bill unless they will allow for difference in price; so Xmas is coming on when business is blocked, I want a settlement quickly. I am aiming at Liptons, the largest grocer in the world, permanently stocking Pure Unadulterated goods, so pray do not send me any as I know you might want to for they would come too late.

Oliver Heaviside, champion of consumer rights!

He wrote to Searle in December 1924, joking about the fall and inviting him and his wife to visit. The four-year freeze in their relationship was over. But soon afterwards he wrote again, saying 'jaundice, so don't expect much'. They called on New Year's Day and found him, impish as always, but frail and yellow with the jaundice. He asked them to get him some handkerchiefs. They took them the next day but could get no answer to their knocking. It seemed as though their friend had decided to shut them out of his life again but two days later a man called at their hotel. He was a doctor. Oliver had been found unconscious in his house by Constable Brock, who straightaway alerted his nieces. They had called two doctors and were considering moving their uncle to a nearby nursing home but he had insisted that nothing be done until he had spoken to his friend Searle.

The doctor who called at the Searles' hotel drove them to Homefield where they found Oliver, conscious but very weak, with Brock and the other doctor. After talking to his friend, Oliver agreed to go to the Mount Stuart Nursing Home and an ambulance was sent for. It was Oliver's first ride in a motor vehicle.

After a few days he was better and the Searles called to have tea with him almost every day until their holiday was over. Mr Tree from the Institution of Electrical Engineers came down from London and for several days joined them for tea. Searle reports that Oliver was full of fun and won the affection of everyone in the nursing home, especially the little ward maid, whom he called the Marchioness. But his body, worn out by years of ill-nourishment and a hundred nagging afflictions, had run its course and he died on 3 February 1925.

Oliver was buried with his parents in Paignton cemetery. The funeral, on 6 February, was attended by family members and Mr Tree. The journals and papers carried glowing obituary notices, and his neighbours and others who had come into contact with the odd resident of Homefield were astonished to read of his worldwide fame. His name appears at the bottom of the gravestone, under those of his parents, and cannot be seen when the grass grows a little – a metaphor, perhaps, for the way his name has been neglected elsewhere. But a curious thing happened in the summer of 2005. Visitors found that the gravestone had been cleaned of years of grime and was now shining white, the area around it had been cleared, and the lettering picked out sharply so that 'Oliver Heaviside F.R.S.' could again be clearly seen by anyone walking past. The person who did the good deed wanted to remain anonymous and the local papers ran 'grave mystery' stories.

Perhaps the grave renovation is a metaphor, too. Just as Heaviside had many well-wishers during his lifetime, there are many now who would like to see him achieve his proper place in the public's affections. The work of great poets and painters is there for all to see and their personal lives are remembered as much for their faults as their virtues. Heaviside was a great artist of a different kind and lived his life as a free spirit. He was always provocative, often amusing, sometimes infuriating but never dull. His legacy is not tidily labelled but you don't have to go to a book or gallery to find it. It is all around us, every day of our lives.

Heaviside's legacy

Electricity lights our homes and streets, cooks our dinners, keeps us in touch with one another and brings entertainment to our living rooms. Almost every aspect of our way of life depends on electrical technology of one kind or another. This technology rests on a solid base of theory. The engineers who design, build and maintain equipment have a well-established set of rules to guide them and a full range of ready-made theoretical tools at their disposal. These rules and tools are, in turn, derived from the laws of physics, discovered and developed by men such as Ampère, Faraday, Thomson and Maxwell, but before the full thrust of scientific knowledge could be put at the service of practical application someone had to bridge the gap between the world of the scholarly scientist and that of the practical-minded engineer.

In the mid-to-late nineteenth century the gap was huge. The practical men had, so far, got on very well with only a tiny smattering of theory and they scorned dilettante academics. It was widely believed by engineers that, except when voltages and currents were steady, electricity behaved in a way that was largely incapable of mathematical description and that the only way to understand it was to get a hands-on feel by practical experience. On the other side, most physicists took very little interest in practical problems. To bridge the gap required a man who knew both worlds – who knew what it was like to grapple, cut and splice cable in the middle of the North Sea, and to make theoretical discoveries through hard and sustained intellectual effort.

In fact, not even these attributes were enough. What gave Heaviside the power to bridge the great chasm was his unique approach to mathematics. To him, symbols and equations were not pale abstractions but components in a vivid picture of the physical world. And it was an engineer's picture – one of forces, stresses, strains and energy flow. Being self-taught and of a fiercely independent turn of mind, he was not bound by custom or convention and strode boldly into territory shunned by orthodox mathematicians. This way, he not only solved many intractable problems, but enabled others to follow by opening up new routes that bypassed the time-honoured obstacles and could be used by people without special training in advanced mathematics.

Perhaps the most extraordinary aspect of Heaviside's achievement is that he spent most of his life writing for an audience that did not yet exist. The practical men didn't understand the mathematics, the physicists admired his mastery of electricity but were perplexed by his emphasis on engineering aspects, and the mathematicians were horrified at the liberties he took with their subject. He took the lead in creating what was, in effect, a completely new discipline – electrical engineering science – with its own language and its own way of looking at the physical world. Eventually the world caught up with him. When the big telecommunications companies began to build research laboratories and universities opened departments of electrical engineering, Heaviside became their guiding spirit. His terms, such as inductance, impedance, distortion and attenuation[1], his generalisation of Ohm's law, his method of vectors, his interpretation of Maxwell's theory, and his analysis of the transmission line became the common currency of the new breed of engineers.

Great credit goes to the influential engineers who began to turn the tide. Among the first in Britain to spread the word among their colleagues were John Perry and his close friend W.E. Ayrton, who, by a pleasing irony, had been a protégé of William Preece. In America Charles P. Steinmetz was followed by E.J. Berg, Vannevar Bush, J.R. Carson, and other men of stature. They did not champion Heaviside out of any tender feelings they may have held for a man whom life had treated harshly, but from hard-headed conviction that his ideas were the right ones. He had not made things easy for them. Perry was one of those who had for years implored him to make his writings easier to follow, and some of the more mathematically inclined engineers even worried about Heaviside's disdain for rigour: Carson complained that his operational calculus, while a godsend, was 'full of holes'. And much of his work was poorly signposted. The way he usually set out his thoughts, especially on circuit theory, was to present a host of specific cases, and what others would regard as the key *teaching* points of the theory were not always set out explicitly.

There were exceptions, notably his introduction to vector analysis. Here, for example, he avoided the laborious algebraic proof of Stokes' theorem that Maxwell and others had used. His derivation is simpler, more direct, and largely dispenses with symbols. It has now become the standard method.[2] And in circuit theory his concept of characteristic impedance, now indispensable to communications engineers, opened up a clear route to the wider understanding of transmission lines. If you terminated a line with the appropriate value of impedance – now called the characteristic impedance – the line would behave as though it were infinitely long: all waves travelling down the line would be absorbed at the receiving end, none reflected. To arrive at this result he worked out the impedance of an infinitely long line with evenly distributed resistance, inductance, capacitance and leakance in a way that was breathtaking in its simplicity and elegance. Suppose the line is already infinitely long. Then, if you add a short extra length the impedance will be unchanged. Apply a little elementary circuit theory and the required formula falls out. What is more, the distortionless condition $L/R = C/G$ is laid bare.[3]

Such pedagogic gems occurred where Heaviside's thinking happened to run along lines comprehensible to those less gifted. Elsewhere, one had to overcome his tendency to 'jump deep double fences', as Fitzgerald had put it. And it was impossible to skim-read Heaviside's papers: you simply had to spend time working carefully through them. But, little by little, his once outrageous ideas found their way into college courses and textbooks, and in the end were so thoroughly absorbed that they became run-of-the-mill.

To give Heaviside *all* the credit for electrical engineering theory would, of course, be absurd: there is plenty to go round. Many others would appear on a list of honour, including his French contemporaries Léon Charles Thévenin, who showed how to represent a complicated circuit by an equivalent simple one, and Aimé Vaschy, who independently advocated increasing inductance in transmission lines shortly after Heaviside had done so. And, in a curious reverse pun, Oliver wrote very little about the heavy side of the subject – electrical power and machines.[4] The technology and theory there are simpler than in electrical communications – the comparison is something like that between the clock of Big Ben and a Cartier wristwatch – but their history is just as fascinating, with even more shenanigans in the scramble for patents and fortunes. Most of the action took place in America. Three of the key figures were Nikola Tesla – like Pupin, a Serbian immigrant – who invented polyphase AC machines and distribution systems; George Westinghouse, the entrepreneur who took up Tesla's inventions; and Charles Proteus Steinmetz, who led research at the Westinghouse Company's rival-turned-collaborator General Electric and started up the Department of Electrical Engineering at Union College. Steinmetz combined prodigious inventive flair – he held about 200 patents – with equally impressive theoretical ability and, as we have seen, was a great admirer of Heaviside's work.

One of Steinmetz's achievements was to get colleagues and students to accept a mathematical device that allowed the theory of DC circuits (where currents were steady) to be extended to AC circuits (where currents oscillated at a fixed frequency). The trick was to replace the ordinary numbers which represented the electrical resistance of each branch in the circuit by complex numbers of the form $X + jY$, where X and Y were ordinary numbers and j (or $-j$) was the square root of -1.* They represented what Heaviside called impedance; the X part was a resistance and the jY part either an inductance or a capacitance, or a combination of the two. For a resistor the component of impedance was just R, for an inductor it was $j\omega L$ and for a capacitor it was $1/j\omega C$, or $-j/\omega C$, where ω was 2π times the current frequency.† One could now apply Ohm's law as if for a DC circuit.[5] When j^2 came up in the calculations you simply replaced it by -1. This way you could work out, for a given alternating voltage, not only the magnitude of the current but also its *phase* – whether it lagged behind the voltage by a fraction of a cycle or led it, and what the fraction was. The scheme worked because the bizarre-seeming symbol j, regarded as an *operator*, had a simple physical meaning – it had the effect of advancing the phase of the current or voltage on which it operated by a quarter of a cycle.[6]

All this fell out of Heaviside's operational calculus. It was a simple special case: when a regular sinusoidally oscillating voltage was applied to a circuit, his differential operator p became $j\omega$, and the rest followed. Heaviside had worked this out long before AC power systems came into commercial use but he never presented the results in a form that power engineers could easily assimilate and it was Steinmetz who took up Heaviside's ideas and demystified complex numbers for them.

The operational calculus itself was a boon to engineers with problems to solve but it had a kind of underground existence, being generally thought too raffish to be

---

* Engineers had a reason for preferring j to the mathematicians' i. See Note 6.
† Heaviside was at pains to point out that this was a bona fide 2π – the angular measure in radians of a complete rotation or cycle – and had nothing to do with the anomalous 4πs that he wanted to eliminate from the equations of electromagnetism.

taught in respectable classrooms. It eventually gained complete respectability but only by taking on extra mathematical machinery and moving the analyst a little further away from the physical world towards one of abstract symbols. Would Heaviside have approved? Probably not, but the system now taught and practised everywhere is essentially his, dressed in fine clothes. The first critical steps were taken in 1916 by a Cambridge friend of Searle's, Thomas John l'Anson Bromwich. He found that he could solve many of Heaviside's operational equations with full mathematical rigour by treating the operator p as a complex number and using the advanced technique of complex integration. Bromwich actually thought Heaviside's own methods were wonderful – with them he could work out solutions mentally without even the need for pencil and paper – but told Oliver he wanted to find a way of 'convincing the purest of pure mathematicians that the p-method rests on sound foundations'. He got no thanks from grouchy Oliver, who replied saying 'I could never stomach your complex integral method', but his paper stimulated other mathematicians to take an interest in Heaviside's operational methods.

In the 1920s, J.R. Carson, a mathematically gifted engineer at The American Telegraph and Telephone Company, came up with a formula for generating Heaviside's p function automatically from the corresponding function of time. The German mathematician Gustav Doetsch brought everything together in the 1930s by combining the approaches of Bromwich and Carson in a complete system with impeccable mathematical credentials. This was the method of so-called Laplace transforms which has become a standard tool for engineers and physicists.[7] The unit step function that Heaviside used all the time and denoted by a bold-faced **1** is now commonly named after him and written as $H(t)$, but many of the people who use it with their Laplace transforms know nothing of the story behind the H.

Heaviside's interpretation of Maxwell's theory of electromagnetism has been enormously influential. Not only did he encapsulate the theory in the four beautiful equations now known everywhere as 'Maxwell's equations', he devised the very language of vectors in which they are expressed. Even more significantly, he showed physicists and engineers how to *think* about electromagnetism: how to use Maxwell's theory to analyse the forces and the fluxes, and the flow of energy in space. What has become the way of the world, when it comes to learning and applying Maxwell's theory, is Heaviside's way. Textbooks on Maxwell's theory, which began to appear in the 1890s, generally bear Heaviside's stamp as surely as they do Maxwell's.[8] And when students of electromagnetism get to grips with the mathematics of their subject they no longer have to contend with the dense undergrowth of anomalous 4πs that Heaviside fought so hard to remove.

The advance of electrical communications in the past hundred years is the greatest leap of technology in humankind's history. Radio, television, radar, mobile phones, the Internet, satellite navigation: each has been turned from ambitious idea into everyday tool at incredible speed. With astounding skill and ingenuity, scientists and engineers have created hundreds of wonderful new devices and techniques, like the thermionic valve, the transistor, the silicon chip, the laser and fibre optics. On the face of it, we have left Heaviside way behind. But in one important sense it was Heaviside who made it all possible. By bridging what had been a great gulf between theory and practice he brought advanced electrical science within the reach of technologists.

He would probably not be all that surprised at modern developments. In his paper 'On electromagnets, etc.' in 1878, he wrote:[9]

> As ... the telephone is sensible to very much more rapid reversals than 1,000 per second, the enormous speeds possible on short lines are easily conceivable, would the action be sufficiently magnified and recorded, so as to appeal to the eye instead of the ear.

It sounds very like cable television. And what are digital radio and television but advanced forms of high-speed telegraph?

How would he wish to be remembered? Heaviside gave us clear guidance on this point in his moving tribute to Maxwell.[10]

> A part of us lives after us, diffused through all humanity – more or less – and all through nature. This is the immortality of the soul.

A part of Heaviside lives today in the work of all who study electricity and put its amazing properties to good use – and in the everyday lives of the far greater number who benefit from their efforts. A splendid legacy indeed, but it will be our loss if we forget the extraordinary man who gave it to us.

Notes

I have tried to tell the story of Heaviside's life simply and directly, putting the reader at his side, seeing the world from his perspective as his life unfolds. Hence the main narrative contains few references to sources and no more historical or technical background than is needed for the story. The Notes attend to these aspects. They also shed some interesting sidelights and I hope that all readers will enjoy browsing through these pages.

To keep clutter to a minimum, dates of papers and letters are generally given by year only, but I have included month, and sometimes day, where they are relevant to the story or where it seems otherwise helpful to do so.

Chapter 1   Do try to be like other people

1. King Street in Camden Town, where Heaviside was born, is now called Plender Street.
2. The description 'Lowest class. Vicious, semi-criminal' was given to the population of an area close to King Street by C. Booth in his study *Life and Labour of the People of London*, published in 1902.
3. The parable about the boy who was beaten with a strap and eaten up by lions is the first paragraph of Volume III of Heaviside's treatise, *Electromagnetic Theory*.
4. The folly of attempting to teach Euclid to schoolboys was one of Heaviside's favourite themes. The passage quoted here is from his article 'The Teaching of Mathematics', published in *Nature* in October 1900 and later included in Chapter 10 of *Electromagnetic Theory*, Volume III.
5. The passage beginning 'I always hated grammar' is from Chapter 4 of Heaviside's *Electromagnetic Theory*, Volume I. The context of this passage, with its plea for simple words, was his advocacy of the name 'mac', after Maxwell, for the unit of inductance. He didn't gather much support. Other Maxwell enthusiasts thought 'mac' disrespectful, a bit like calling the Queen 'Vicky'. The name eventually adopted for the unit of inductance was the henry, after Joseph Henry, the great American contemporary of Faraday.
6. Whatever spurious kudos Wheatstone gained from the Wheatstone Bridge – actually invented by Samuel Christie – is balanced by the genuine credit he lost when his ingenious new cipher system was named after Lyon Playfair, the chemist turned politician who had promoted its use by the British government. The Playfair code was easy to use but more secure than earlier codes because letters were coded in pairs.

7. Heaviside wrote the account of his early life, beginning 'I was born and lived 13 years in a very mean street in London', in a letter of June 1897 to his friend George Francis Fitzgerald.
8. Heaviside used the parable beginning 'More than a third of a century ago, in the library of an ancient town' to introduce a review of the book *Vector Analysis*, by Gibbs' former student E.B. Wilson. The review was published by *The Electrician* in March 1902 and reprinted as Appendix K of *Electromagnetic Theory*, Volume III. The parable was Oliver's way of expressing his exasperation with quaternions. (When the knowledge-seeking youth finds books on quaternions he studies them eagerly but finds them incomprehensible and dies of disappointment.)

Chapter 2   Seventy words a minute

1. The French mathematician Henri Poincaré was fond of relating how a colleague who had made a long study of Maxwell's *Treatise on Electricity and Magnetism* had told him: 'I understand everything in this book, except what is meant by a charged sphere.' Today we know from quantum electrodynamics that electrical forces result from electrons and quarks emitting and absorbing photons but, as Richard Feynman has remarked, nobody knows *why* Nature works in this way.
2. Schilling's scheme may have been the first practical electromagnetic telegraph but the English inventor Francis Ronalds had produced a prototype *electrostatic* telegraph in 1816 and offered it to the Admiralty. In one of the most extreme and graceless examples of official myopia, their Lordships replied that 'telegraphs of any kind are now wholly unnecessary'. After years of bitter war with France the country was glad to be at peace and the Government was not inclined to look to the future. Ronalds gave up his telegraphic investigations shortly after receiving this snub and so never took advantage of Oersted's discovery of the magnetic effect of electric currents. His many other inventions included fire alarms and machines for making perspective drawings but he is now best remembered for his pioneering work on photographic and electrical methods in meteorology. At the age of 83, two years before he died, he was rewarded with a knighthood.
3. William Thomson was to all intents and purposes a scientific consultant to the Atlantic Telegraph Company but was not formally contracted as such. He was instead appointed to the board of directors, all of whom were unpaid. Like the other directors, Thomson stood to gain financially if the venture succeeded, and he also had the prospect of income from the company's use of his patents.
4. The comment that Whitehouse sent a stroke of lightning through the cable when only a spark was needed was made by P.B. McDonald in his book *A Saga of the Seas, The Story of Cyrus W. Field and the Laying of the First Atlantic Cable*.

Chapter 3 Waiting for *Caroline*

1. Heaviside wrote this account of his first sight of Maxwell's *Treatise* in a letter to the French electrical engineer Joseph Bethenod in February 1918. Bethenod, a great admirer of Heaviside, included it (translated into French) in an obituary notice for Heaviside published in *Annales des Postes Télégraphes* in 1925. I am indebted for its inclusion here to Paul Nahin, who quoted it (after retranslation into English) in his biography *Oliver Heaviside: Sage in Solitude*. The perils of double translation are well known: there is a popular story of 'out of sight, out of mind' becoming 'invisible idiot' when translated into Russian and back. But in this instance both Bethenod and Nahin were careful to translate literally, so we can be confident that the quoted words are close to Heaviside's own.
2. Rollo Appleyard describes the time of Oliver's experiments in Beckett's shop as 'perhaps the happiest day of his existence' in his book *Pioneers of Electrical Communication*.
3. Heaviside's paper 'On Duplex Telegraphy, Part I', was published in the *Philosophical Magazine* in June 1873. Part II came out in January 1876. They were reprinted as Articles 7 and 8 in the collected *Electrical Papers*, Volume I.
4. Preece's complaint to Culley of Heaviside's 'most pretentious and impudent paper', and Culley's reply, are reported by E.C. Baker in his biography *Sir William Preece F.R.S.*

Chapter 4 Old Teufelsdröckh

1. Heaviside compared himself to 'old Teufelsdröckh in his garret' in a letter of July 1908 to Joseph Larmor.
2. Rollo Appleyard reports in his book *Pioneers of Electrical Communication* that Heaviside was a good gymnast who took the pastime seriously enough to list the exercises he performed and record his body measurements.
3. In the same book, Appleyard quotes Heaviside's letter containing the passage beginning 'In old days I went to concerts'. He does not name the friend to whom the letter was written or give the date.
4. There is a rough scrap of musical notation in the draft of a letter Heaviside wrote to his French friend Joseph Bethenod in 1921 – he asks him whether the tune of 'Pop Goes The Weasel' is known in France and sets out the tune like a graph, with short horizontal lines for the notes, the height of each line representing the pitch of the note relative to the others. His purpose here was just to convey a simple tune and we don't know how he proposed to represent chords, rhythm and key. The same letter is referred to in Note 21 to Chapter 10.
5. The abbreviations T and T′ had their origin in a lively three-way correspondence conducted by the three Scottish physicists William Thomson, P.G. Tait and James Clerk Maxwell. They wrote on postcards and to get as much as possible on a card they abbreviated words wherever they could, starting with names. Thomson was T, Tait was T′ and Maxwell was dp/dt (from the equation dp/dt = JCM in Tait's

book on thermodynamics). In their code, Hermann Helmholtz was H^2 and John Tyndall was T''. Tait did not have a high opinion of Tyndall and, in his cutting way, used to say that T'' indicated a second-order quantity.

6. The passage beginning 'For my part I always admired the old-fashioned term "natural philosopher"' is from the Introduction to Volume I of his *Electromagnetic Theory*. The terms physicist and biologist had not yet become accepted and, in the same paragraph, Heaviside, tongue in cheek, suggests 'materialist' and 'organist' as alternatives. The paragraph finishes with a typical off-beat flourish.

    > How is it possible to be a natural philosopher when a Salvation Army band is performing outside; joyously it may be, but not melodiously? But I would not disparage their work; it may be far more important than his.

7. Heaviside's paper 'On the Extra Current' was published in the *Philosophical Magazine* in August 1876 and reprinted as Article 14 in *Electrical Papers*, Volume I.

8. Heaviside explained why information could be sent faster eastwards than westwards on the Anglo-Danish cable in his paper 'On the Speed of Signalling through Heterogeneous Telegraph Circuits', published by the *Philosophical Magazine* in March 1877 and reprinted as Article 15 of *Electrical Papers*, Volume I.

9. Henry Cavendish (1731–1810), the English chemist and physicist, was a great genius. In one experiment he estimated the density of the earth within 2% of its true value. He was also a great recluse who communicated with his servants by written notes. Sometimes he published his findings but more often did not. Much of his electrical work was unknown to other scientists until Maxwell edited and published it in the 1870s.

10. Heaviside showed how to analyse AC circuits in his paper 'On Electromagnets, etc.', published in the *Journal of the Society of Telegraph Engineers* in 1878 and reprinted as Article 17 of *Electrical Papers*, Volume I.

Chapter 5   Good old Maxwell!

1. The rent on Heaviside's parents' house at 3 St Augustine's Road was £45 per year and their total household expenditure was probably at least four times that figure. All four brothers helped financially but Oliver's contribution would have been much smaller than those of Arthur and Charles.

2. *The Electrician* was owned by the Eastern Telegraph Company, whose chairman was Sir John Pender. He was a Manchester textile merchant who had spotted early the prodigious possibilities of the telegraph and had taken up his new interest with vigour, becoming chairman not only of the E.T.C. but also of the Telegraph Construction and Maintenance Company, which made and laid many of the world's great under-sea cables. This was not all; he found time for politics and became a Liberal M.P. With so many interests he was a hands-off proprietor of *The Electrician* and the quarrels that Biggs and later editors had with the editorial board were probably not with him but with his acolytes. One of the

ousted editors, A.P. Trotter, wrote to Heaviside in 1899 that he had not forgiven the 'Pender parasites who drove me from *The Electrician*'.

3. *The Electrician* published Heaviside's paper 'The Earth as a Return Conductor' in November 1882. The paper was reprinted as Article 23 of *Electrical Papers*, Volume I.

4. Heaviside made his joke about the divine creation of Ohm's law in one of a series of papers on 'The Energy of the Electric Current', published in *The Electrician* in 1883. When the paper was republished as Section 5c of Article 27 of *Electrical Papers*, Volume I, the offending passage was omitted.

5. The tribute to Maxwell beginning 'A part of us lives after us, diffused through all humanity' was not included in any of Heaviside's publications. It is quoted by Rollo Appleyard in his book *Pioneers of Electrical Communication*.

6. Airy's disparaging comment on Faraday's lines of force is quoted by J.J. Thomson in his essay 'James Clerk Maxwell' in *James Clerk Maxwell: A Commemoration Volume* which was published in 1931 by the Cambridge University Press to celebrate Maxwell's centenary.

7. Interested readers can find a non-technical description of Maxwell's spinning cells model in my book *The Man Who Changed Everything: The Life of James Clerk Maxwell*, and a guided mathematical study of the model in Thomas K. Simpson's *Maxwell on the Electromagnetic Field*.

8. The passage beginning 'Maxwell, like every other pioneer who does not live to explore the country he opened out' is from a review by George Francis Fitzgerald of Heaviside's *Electrical Papers*. The review was published by *The Electrician* in August 1893.

9. Heaviside introduced vector analysis in his series of papers 'The Relations between Magnetic Force and Electric Current', published in *The Electrician* in 1882 and reprinted as Article 24 of *Electrical Papers*, Volume I.

10. The vectors **E** and **H** are not forces in the ordinary sense but rather the *intensities* of the electric and magnetic fields at our arbitrary point. But they are the forces that would act on a unit charge or a unit magnetic pole if either were placed at the point and Heaviside, who much preferred vivid descriptions to abstract ones, always called them forces.

The constants μ and ε are the ratios of magnetic and electric flux densities to their respective forces. The curl equations can contain various arrangements of μ and ε according to the system of units used. And, today, the magnetic flux density vector $\mathbf{B} = \mu\mathbf{H}$ is generally used rather than the force vector **H**. Heaviside favoured the symmetrical arrangement shown in the narrative.

In a deeper sense, there is only one fundamental electromagnetic constant – the speed of light, c – and its value has to be measured by experiment. By Maxwell's theory, c is equal to $1/\sqrt{(\mu\varepsilon)}$. This means that either μ or ε can be arbitrarily fixed to suit a particular system of units, but when this is done the other is automatically fixed, too, because $\mu\varepsilon$ must always be equal to $1/c^2$. In the system of units most commonly used in Heaviside's time, μ was set at 1.

I have, for simplicity, omitted the suffix 0 generally given to μ and ε when, as here, they represent the ratios of flux density to force in a vacuum, technically known as the permeability and permittivity of free space. The same symbols with suffix *r* are used to represent the *relative* permeability and permittivity of particular substances compared with those of a vacuum.

11. The inverse square law for electric and magnetic forces can be visualised by taking as an analogy the outward flow of an incompressible fluid from a point source. The vector in this case is the velocity of flow and its divergence is zero everywhere but the source because the fluid is incompressible. The amount of fluid emerging per unit time from any sphere centred on the point source will be the same, whatever the size of the sphere. It follows that the fluid moves outwards at a rate inversely proportional to the square of its distance from the source, and, by analogy, that electrical and magnetic forces follow the same law.

12. Heaviside published his reformulation of Maxwell's theory in the first half of the long series of papers 'Electromagnetic Induction and its Propagation', which started to appear in *The Electrician* in 1885. The first half of the series was reprinted as Article 30 of *Electrical Papers*, Volume I, and the second as Article 35 in Volume II.

13. The terms curl and divergence were coined by Maxwell. (His suggestion was actually convergence but it has been adopted in negative form, in line with Heaviside's preference.) Maxwell also proposed the term slope for the vector representing the rate of change in space of a scalar field and this, too, has been universally adopted, although in slightly more formal guise, as gradient, or grad.

14. The potential function was the inspiration of Pierre Simon de Laplace, author of *La mécanique céleste*, the standard work on mathematical astronomy. An example of use of potentials in the solution of a difficult problem in astronomy is Maxwell's analysis of Saturn's rings. Forming the equations of motion in terms of the rings' gravitational potential, he showed that the rings must consist of many discrete solid bodies orbiting independently, a prediction borne out by flypast pictures from the Voyager and Cassini spacecraft.

15. Richard Feynman's remark about potentials being the fundamental quantities in quantum electrodynamics is from Volume 2 of *Lectures in Physics*.

Chapter 6   Making waves

1. Heaviside gave the formula for energy flow in Section 4 of the first half of the series 'Electromagnetic Induction and its Propagation', published in *The Electrician* in January 1886 and reprinted in Article 30 of *Electrical Papers*, Volume I.

2. Heaviside first gave the energy flow formula – in words, not symbols, and for special case of energy transmission within an iron core inside a coil – in June 1884. It was in Section 19 of 'Induction of Currents in Cores', first published in *The Electrician* and reprinted as Article 28 of *Electrical Papers*, Volume I.

3. Heaviside made the reference to 'This remarkable formula' in the second chapter of *Electromagnetic Theory*, Volume I. It appears in the section headed 'Electromagnetic application. Medium at rest. The Poynting flux'.

4. The American Henry Rowland had shown in 1875 that an electrically charged spinning disc creates a magnetic field. In 1889, Heaviside himself was the first to give the correct formula for the mechanical force on an electric charge moving in a magnetic field. His result was hidden away in the middle of a long paper and the force has become named after the Dutch physicist Hendrik Anton Lorentz, who published the same result a few years later. Heaviside's paper also contained the first appearance of what became known as the 'relativistic factor'; see Note 2 to Chapter 10. On electric charges and forces, see also Note 1 to Chapter 2.

5. The quotation 'The human mind is seldom satisfied, and is certainly never exercising its highest functions, when it is doing the work of a calculating machine' is from Maxwell's presidential address to Section A of the British Association for the Advancement of Science in 1870.

6. The comparison of Lodge's book to a factory was made by the French physicist and philosopher Pierre Duhem. He was representing the view, widely held on the continent, that British physicists relied too heavily on mechanical models.

7. Fitzgerald's short paper 'On a Method of Producing Electromagnetic Disturbances of Comparatively Short Wavelengths' was published in *The Electrician* in 1883 and is included in *The Scientific Writings of the Late George Francis Fitzgerald*, edited by Joseph Larmor, published in 1902.

8. When Lodge published his collected papers on the topic in a book *Lightning Conductors and Lightning Guards* in 1892, he included the praise of Heaviside's work but, having by then become a friend of Oliver, he edited out the adjectives 'eccentric' and 'repellent'.

9. Heaviside wrote to Lodge, saying 'I looked upon your 2nd lecture when I read it as a special kind of providence' in September 1888.

10. Rollo Appleyard describes Hertz as going to Berlin 'to acquire the stride of the giants' in *Pioneers of Electrical Communication*.

11. Hertz's diary entry in which he says he 'hit upon the solution of the electromagnetic problem this morning' was written in May 1884. Although we don't know what particular problem this was, the entry clearly demonstrates Hertz's deep interest in theoretical aspects of his subject.

12. Heaviside made his comment about Helmholtz's theory being 'Maxwell's run mad' in a letter of June 1892 to Lodge.

13. The remark that news of Hertz's discoveries reached Germany by way of England was made in an obituary of Hertz published in *Nature* in 1894.

General note

A subject that absorbed the attention of many late nineteenth-century physicists, among them Lodge and Fitzgerald, was the composition of the aether – the hypothetical substance, pervading all space, that was thought to be the medium by which light and other electromagnetic waves were transmitted. Both Lodge and Fitzgerald spent much time making imaginary mechanical models of the aether. Lodge's favourite model had a rack-and-pinion mechanism and Fitzgerald's had pulleys and bands.

They performed a similar function to Maxwell's spinning cells model of 1861 but whereas Maxwell had abandoned his model once he had found a way to build his theory without it, Lodge and Fitzgerald held on to theirs. Heaviside took little interest in mechanical models. Like everyone at the time, he thought an aether of some kind was necessary to store electromagnetic energy and to transmit waves, but he followed Maxwell in keeping an open mind about its structure.

The aether of nineteenth-century physics needed to operate in absolute space and time, but in 1905 Albert Einstein demolished those notions in his special theory of relativity. There he showed that all measures of space and time were relative and that the only constant quantity was the speed of light. There was no longer a home, or need, for the aether; in their new guise, space and time had themselves taken on the role.

So all the efforts of the aether model builders came to nothing in the end. Or did they? The keystone of Maxwell's enduring theory, the displacement current, had its origin in the idea that the spinning cells in his construction-kit model could be springy. And although the dozens of models proposed by Lodge, Fitzgerald, Thomson and others have no direct relevance today they served in their time as powerful stimuli for thought and so contributed to the general development of physical science. The models of Lodge, Fitzgerald and their contemporaries are compellingly described by Bruce J. Hunt in *The Maxwellians*, and E.T. Whittaker gives a full treatment of the topic in his splendid book *A History of the Theories of Aether and Electricity*.

In one sense the notion of an all-pervading aether lives on. Physicists today believe that what we loosely think of as empty space is awash with virtual particles popping in and out of existence.

Chapter 7   Into battle

1. The passage beginning 'Oliver Heaviside has the faults of extreme concentration of thought' is from a review by Fitzgerald of Heaviside's *Electrical Papers*, published by *The Electrician* in August 1893. The same review contained the description, quoted in Chapter 5, of Heaviside clearing away Maxwell's 'debris'.
2. An obituary in *The Electrician* in 1900 records David Hughes' birthplace as London but more recent research suggests it may have been Corwen, in Denbighshire, North Wales. He emigrated to America with his parents when he was seven years old.
3. Heaviside expressed his exasperation with Hughes' experimental results in the first of two 'Notes on the Self-Induction of Wires', published by *The Electrician* on 23 April 1886 and republished in Article 33 of *Electrical Papers*, Volume II.
4. Hughes' prompt response to Heaviside's criticism appeared in the next issue of *The Electrician*, on 30 April 1886.
5. In his turn, Heaviside responded to Hughes in a second 'Note on the Self-Induction of Wires', published in *The Electrician* the following week, on 7 May 1886, and republished in Article 33 of *Electrical Papers*, Volume II. The well-publicised discussions of David Hughes' experiments helped to bring home to electrical engineers that they could no longer manage without mathematics. Yet Hughes himself didn't believe in the skin effect and never really understood

what was going on. Dominic Jordan skilfully unravels the threads of this interesting but convoluted affair in his paper 'D.E. Hughes, Self-Induction and the Skin-Effect', listed in the Bibliography.

6. Heaviside's account of his investigation into the compensating effects of resistances and leaks can be found towards the end of Chapter 4 of *Electromagnetic Theory*, Volume I. I have, for simplicity, omitted from the main narrative one subtle aspect of Heaviside's analysis. He pointed out that the compensation is only approximate for finite resistance and leakage elements but becomes exact when they are infinitely small, as in a real line with continuously distributed resistance and leakage.

7. In Heaviside's formula I have, for clarity, used the symbols generally used today. He used S, rather than C, for capacitance and K, rather than G, for leakance.

8. Preece's remarks challenging Maxwell's work on self-induction were made in 1889 and are quoted by Dominic Jordan in his excellent article 'The Adoption of Self-induction by Telephony 1886–1889', listed in the Bibliography. This article is also the source of the wonderfully epigrammatic line 'Heaviside's whole life can be read as a kind of allegory on the ultimate triumph of virtue', which I have quoted at the start of the book.

9. Preece expounded his KR law in the article 'On the Limiting Distance of Speech by Telephone', published in the *Proceedings of the Royal Society*, vol. 42, pp. 152–158, 3 March 1887.

10. Thompson's biting put-down of Preece is quoted by Jordan (see Note 8 above).

11. Heaviside asked his satirical question about whether the current knows that it is going to Edinburgh in Section 42 of his long series 'Electromagnetic Induction and its Propagation'. This Section was published in *The Electrician* in July 1887 and later appeared as part of Article 35 of *Electrical Papers*, Volume II.

12. Heaviside's series of six papers 'On Electromagnetic Waves' was published by the *Philosophical Magazine* between February and December 1888. The full title was 'On Electromagnetic Waves, especially in relation to the Vorticity of the Impressed Forces; and the Forced Vibrations of Electromagnetic Systems'. It was reprinted as Article 43 of *Electrical Papers*, Volume II.

Chapter 8   Self-induction's in the air

1. Most of the audience at Bath would have been familiar with the story from the Old Testament book of Numbers, Chapter 22. Balaam was hired by Balak to curse the Israelites, but God told him not to carry out this commission but rather to prophesy that the Israelites would prevail over all the indigenous tribes and settle in Canaan. Balak offered to raise his fee but Balaam refused all payment and chose instead to obey God's instruction. The prophecies came true.

2. The passage in which Heaviside asks the question 'Is self-induction played out?', and proceeds to answer it, is from his paper 'Practice versus Theory – Electromagnetic Waves', published in *The Electrician* on 19 October 1888 and later as Article 46 of *Electrical Papers*, Volume II.

3. The Society of Telegraph Engineers had broadened its title and scope in 1881 when it became the Society of Telegraph Engineers and Electricians. In 1889 it changed its name again to the Institution of Electrical Engineers, setting itself on a par with the established Institutions of Civil and Mechanical Engineers. In 2006 it merged with the Institution of Incorporated Engineers to form the Institution of Engineering and Technology.

4. William Thomson's Presidential Address to the Institution of Electrical Engineers, in which he praised Heaviside's work, was delivered on 10 January 1889.

5. Heaviside responded with his letter thanking Thomson on 18 January 1889.

6. On the same day, 18 January 1889, *The Electrical Engineer (London)* published C.H.W. Biggs' editorial, in which he welcomed Thomson's eulogy of Heaviside's work and took an enigmatic swipe at Preece. Ever one to see fair play, Biggs came to Preece's *defence* later in the year when the eminent man had been misquoted in the popular press about using electricity for criminal executions. I am obliged to Paul Nahin for mentioning this incident in *Oliver Heaviside: Sage in Solitude*.

7. W.H. Snell, the *Electrician* editor who had stopped Oliver's papers in 1887, became ill the following year and his assistant, W.H. Bond, took on most of the work. Snell died in March 1890, aged only 31, and Bond stayed on as assistant when Trotter was appointed editor.

8. Trotter registered his objection to contentious discussions in a letter to Heaviside dated 29 October 1890.

9. Heaviside's reply, making the extraordinary claim that he never went in for contentious discussions, was sent, almost by return of post, on 1 November 1890.

10. The book *Electromagnetic Waves* was never produced commercially because its contents were reprinted as Article 43 of *Electrical Papers*, Volume II, but the Institution of Engineering and Technology (formerly the Institution of Electrical Engineers) holds a copy in its library. The book includes an extra article, 'The General Solution of Maxwell's Electromagnetic Equations', originally published by the *Philosophical Magazine* in January 1889 and reprinted as Article 44 of *Electrical Papers*, Volume II.

11. Heaviside gave the formula for the distortionless transmission line in Section 40 of his long series 'Electromagnetic Induction and its Propagation', published in *The Electrician* in June 1887 and republished in Article 35 of *Electrical Papers*, Volume II. He gave the formula again in the paper 'On Resistance and Conductance Operators ...' (for full title see Note 8 to Chapter 9), published by the *Philosophical Magazine* in December 1887 and reprinted as Article 42 of *Electrical Papers*, Volume II.

12. Heaviside wrote the passage beginning 'The mathematics was reduced, in the main, to simple algebra' in the Introduction to Chapter I of *Electromagnetic Theory*, Volume I.

13. The evidence that Heaviside discussed the idea of inductive coils with Thomson is circumstantial but compelling. In *Oliver Heaviside: Sage in Solitude*, Paul Nahin quotes the following letter from the telephone engineer Alfred Rosling Bennett, published in *The Electrician* in April 1917.

It is very probable that the idea of such coils originated with Heaviside; but, as I have said on other occasions, Lord Kelvin, then Sir William Thomson, in discussing Heaviside's investigations with me in 1888, said that the practical effect could probably be given to them by intercalating suitable coils at intervals in long telephone lines, and he spoke, as I then thought, as if the notion was his own. But he also led me to understand that he was, or had been, in correspondence with Heaviside – of whose labours he spoke in the highest terms – so that he may have been only repeating something communicated by the latter; although, as far as I can make out, the first public mention of coils by that pioneer was in 1893 ...

14. Hertz told Heaviside in a letter of March 1889 that if Maxwell had lived 'he would have acknowledged the superiority of your methods'. Later in the same letter is a striking illustration of Hertz's modesty and generosity of spirit. Continuing the theme of what Maxwell would have thought of later developments, had he lived, Hertz said 'I think he would have more joy in my experiments and would have had more reason to be proud in their result than I can have.'

15. Heaviside congratulated Hertz on giving 'a death blow' to the concept of action at a distance (and hence to the theories of Helmholtz and others which relied on this concept) in a letter of July 1889. Had he known how greatly Hertz revered his mentor, perhaps Heaviside would not have rubbed salt so forcefully in his friend's wound. But with Oliver one can never be sure.

16. Hertz warned Heaviside that his writing was 'a little obscure for ordinary men' in a letter of May 1889.

Chapter 9   Uncle Olly

1. Beatrice's recollection of merry times in the stock-room is reported by G.F.C. Searle in 'Oliver Heaviside – A Personal Sketch', published in *The Heaviside Centenary Volume*.

2. Heaviside sent his first reply to Lodge about candidature for Fellowship of the Royal Society on 30 January 1891. The speed with which the correspondence was conducted is remarkable, and a testimony to the efficiency of the mail trains. Oliver would have received Lodge's letter from Liverpool on the 28th. He must have sent his long draft reply the same day to brother Arthur in Newcastle, received Arthur's comments by return of post on the 30th and sent the revised reply to Lodge the same day.

3. Lodge responded promptly and Heaviside's second letter on the candidature question was posted on 2 February.

4. There were 68 candidates for Fellowship of the Royal Society in 1891. Heaviside was one of 15 elected. Among the other successful candidates was Silvanus P. Thompson, who was having his third go, and among the unsuccessful ones was Joseph Larmor, who was elected the following year.

5. Once again, Heaviside was quick to react. He sent his complaint about the Royal Society's 'Habeas Corpus' to Lodge on 6 June 1891, the day after receiving his copy of the Society's Statutes.

6. Heaviside included this invitation to mathematicians to share his fascination with operational methods in Part I of his series 'On Operators in Physical Mathematics', published in the *Proceedings of The Royal Society*, vol. 52, pp. 504–529, in February 1893.

7. Harold Jeffreys praised the practical convenience of the operational method in his book *Operational Methods in Mathematical Physics*, published in 1927.

8. The idea of generalising Ohm's law came into many of Heaviside's early papers in the context of solving particular problems, but in December 1887 he set out his method systematically in a paper with the accurate but cumbersome title 'On Resistance and Conductance Operators and their Derivatives, Inductance, and Permittance, especially in connection with Electric and Magnetic Energy', published in the *Philosophical Magazine* and reprinted as Article 42 in *Electrical Papers*, Volume II. (Permittance was his preferred word for capacitance.)

9. I have simplified things a little here, to avoid bogging the reader down in terminology. Most of the time, Heaviside called his $Z(p)$ function the resistance operator of the circuit, reserving the term impedance for the special case of AC circuits where voltage and current oscillated sinusoidally at a fixed frequency. However, he did sometimes use impedance in the more general sense. Today we follow Heaviside by using impedance, represented by his symbol Z, for the complex ratio of voltage to current in AC circuits and, where appropriate, use the term *generalised impedance* for the broader case.

10. Heaviside often used the operational calculus to work out the response of an electrical circuit from the time that a voltage source was connected to it. All such analyses employed his unit step function, which had the value 0 up to time zero and the value 1 thereafter. This function is now often written $H(t)$ in his honour, but he wrote it as a bold-face **1**. The voltage of the source could be any function of time, $f(t)$, but, because it was switched into the circuit at time $t = 0$, the input voltage to the circuit was, in Heaviside's notation, $f(t)$**1**. The presence of $H(t)$, or **1**, in the function being operated on is crucial to Heaviside's method. It ensures that $1/p$ is the true inverse of p – in other words that, while p represents differentiation, $1/p$ represents the operation of integration up to time t. But Heaviside, in his cavalier fashion, did not always explicitly write in the **1** in his equations. It was always implied, and he allowed for it in his calculations, but this aversion to rigorous notation could lead followers of his method astray if they failed to make sure to have a **1** in the function they were operating on. Ido Yavetz gives a sympathetic and illuminating appraisal of this and other aspects of Heaviside's operational calculus in his book *From Obscurity to Enigma: The Work of Oliver Heaviside, 1872–1889*.

11. Heaviside made his statement that the square root of a differential operator is intrinsic to physics in Chapter 7 of Volume II of *Electromagnetic Theory*. He was right. It occurs not only in electrical transmission lines but in many other physical systems that are represented by partial differential equations. These are equations describing the rates at which a physical quantity varies with each of two or more variables. Those for a transmission line, for example, describe the variations of voltage and current both with time and with distance along the line.

12. Heaviside found the relation $\sqrt{p1} = 1/\sqrt{(\pi t)}$ by the 'experimental' method that infuriated orthodox mathematicians. Having first solved a transmission line problem by non-operational means, using Fourier series, he then wrote the equations in operational form and observed that the relation had to be true for the operational method to give the right answer, adding breezily at the end: 'The above is only one way in a thousand'. The relation is closely linked to the curious result $(-1/2)! = \sqrt{\pi}$, which was already known to mathematicians. (Heaviside liked to use the factorial function $n! = 1 \times 2 \times 3 \times \ldots n$ outside its proper domain of integers, thereby causing more irritation to orthodox mathematicians. The same result is more conventionally expressed as $\Gamma(1/2) = \sqrt{\pi}$, where Γ is Euler's gamma function; here it comes from a straightforward integral.) Heaviside blazes his way through the mathematics in Chapter 7 of *Electromagnetic Theory*, Volume II.

13. The passage concluding 'You have first to find out what there is to find out' is part of a polemical section, headed 'Mathematics is an Experimental Science', with which Heaviside began Chapter 5 of *Electromagnetic Theory*, Volume II.

14. Fitzgerald's letter comparing Heaviside's earnings to those of a hodman was written on 15 February 1894, nine days after Fitzgerald, Lodge and Perry sent their joint letter. There had been at least one exchange of letters between Heaviside and Fitzgerald in the meanwhile.

15. Heaviside's letter to Fitzgerald, finally refusing the Royal Society's offer of financial help, was written a week later, on 22 February.

16. It was Gibbs who introduced the notation now used for vector and scalar products (see Note 17 below); $\mathbf{a} \times \mathbf{b}$ indicates a vector product and $\mathbf{a.b}$ a scalar product. Heaviside wrote V$\mathbf{ab}$ for the vector product and $\mathbf{ab}$ for the scalar. But the idea of using bold-face letters for vectors was Heaviside's.

17. Hamilton thought that his symbolic scheme of opposite rotations ($ij = k$ but $ji = -k$) would be just the thing to represent physical relationships like that between electricity and magnetism; for example, it was well known that if you reversed the *linear* direction of the current in a straight wire the circular magnetic field around it would also reverse, but in a *rotatory* sense, from, say, clockwise to anticlockwise. But there was a flaw in the scheme. He used i, j and k to represent 90° rotations about three mutually perpendicular axes and the same symbols to represent unit vectors in the x, y and z directions. This led to some strange properties; for example, the square of a vector was negative. Heaviside and Gibbs overcame the difficulty by distinguishing between so-called vector and scalar products. In modern notation (see Note 16 above) bold-faced letters $\mathbf{i}$, $\mathbf{j}$ and $\mathbf{k}$ represent the three unit vectors, and rotations are defined not by $\mathbf{i}$, $\mathbf{j}$ and $\mathbf{k}$ themselves but by their vector products:

$$\mathbf{i} \times \mathbf{j} = \mathbf{k} \quad \mathbf{j} \times \mathbf{k} = \mathbf{i} \quad \mathbf{k} \times \mathbf{i} = \mathbf{j} \qquad \mathbf{i} \times \mathbf{i} = \mathbf{j} \times \mathbf{j} = \mathbf{k} \times \mathbf{k} = 0$$

whereas, in the scalar products:

$$\mathbf{i.i} = \mathbf{j.j} = \mathbf{k.k} = 1 \quad \text{and} \quad \mathbf{i.j} = \mathbf{j.k} = \mathbf{k.i} = 0$$

Problem solved – the square of a vector was no longer negative – but Tait saw this as a desecration of Hamilton's beautiful scheme $i^2 = j^2 = k^2 = ijk = -1$.

18.   As already noted (see Note 13 to Chapter 5), the terms that Maxwell suggested for the properties of vector fields were curl and convergence. Both have been generally adopted, except that convergence has been replaced by its negative form divergence (div for short). Maxwell also suggested the term slope for the rate of change in space of a scalar field and this, too, has been adopted, although in modified form, as gradient (grad for short).

19.   Tait expressed his disappointment at how little progress had been made with the development of quaternions in the Preface to the third edition of his book *An Elementary Treatise on Quaternions*, published in 1890. He had brought out the first edition as a keen young author twenty-three years earlier but, to his dismay, there had, indeed, been virtually no progress since then: very few of his fellow scientists had shown any interest in the subject.

20.   Heaviside made many disrespectful comments about Tait's fixation with quaternions. Those quoted here are from Chapter 3 of *Electromagnetic Theory*, Volume I. Tait would have seen them first when published in article form in *The Electrician*.

21.   Tait trumpeted Knott's 'exposure' of the vector systems of Heaviside and Gibbs in a long letter of January 1893 to *Nature*. Tait and Knott were colleagues at Edinburgh University, where Tait had been Professor of Natural Philosophy since 1860. The electors could have chosen rival candidate Maxwell, who instead went to King's College, London, from where he published his two great papers on electromagnetism in the early 1860s.

22.   *Nature* published Heaviside's response to Tait's letter in April 1893.

23.   Heaviside made his comment 'it is a long way to America' in a letter to *The Electrician* in October 1896.

24.   Perry reported a 'triumph' for Heaviside's operational calculus in his paper 'On the Age of the Earth', published in *Nature* in January 1895.

25.   Heaviside's letter to Fitzgerald, containing the observation on evolution 'Wonderful how things have worked out. If it wasn't true, no one could believe it' was written in July 1897. He did make one reference to evolution in his published papers. Nonchalantly drawing on Harriet Beecher Stowe's *Uncle Tom's Cabin*, he wrote:

> As for the origin of life upon this planet, the only reasonable view seems to me to be Topsy's theory. She was a true philosopher, and 'she spekt she growed'.

This passage can be found about halfway through Chapter 5 of *Electromagnetic Theory*, Volume II.

Chapter 10   Country life

1.   Heaviside wrote to Fitzgerald about his move to Newton Abbot in May 1897.

2.   Searle had written to point out an error in one of Heaviside's papers. Oliver was grateful and made the correction. The subject was how the shape of a spherical

charge changes when it moves – it becomes an oblate spheroid, squashed in the direction of motion by the factor $\sqrt{(1 - v^2/c^2)}$. Heaviside had already shown that the field of a moving point charge would be squashed in the same way. His first reference to what later became known as the 'relativistic factor' was in the paper 'On the Electromagnetic Effects due to a Motion of Electrification through a Dielectric', published in the *Philosophical Magazine* in April 1889 and reprinted as Article 50 of the collected *Electrical Papers*, Volume II.

3. Michelson and Morley carried out their famous experiment in Cleveland, Ohio, in 1887. They wanted to measure the 'aether drift' – the motion of the earth through the substance called the aether which was thought to permeate all space and to be the medium by which light waves were transmitted. They set out to measure the difference in speed between the two parts of a light beam split at right angles. For this they used Michelson's interferometer – an instrument which used the tiny wavelengths of light itself as measurement units and so made possible a degree of accuracy previously unthought of. To their consternation the speed of light in both directions was identical. This was a great disappointment to the experimenters and at first the scientific community saw the experiment as no more than another failed attempt to measure the aether drift. Michelson himself seldom mentioned his result and never recognised its immense significance. But others, including Fitzgerald and Lorentz, began to see that here was some new and important evidence and put forward ideas to account for it.

    Einstein solved the problem in 1905 by turning it inside out. In his special theory of relativity he took the constancy of the speed of light as a starting-point and worked out the consequences. He showed that there were no absolute measures of space or time – all observers in uniform relative motion measured them differently and all their measurements were equally valid. What of the aether? It would need to operate in absolute space and time and when Einstein demolished those there was no longer a home, or need, for the aether. As already observed in the General Note on Chapter 6, space and time, in their new guise, had themselves taken on the role.

4. In Einstein's interpretation it was not that material bodies contracted as they approached the speed of light, but rather that measures of space itself, and time, were different for all observers in relative motion to one another.

5. Heaviside's letter to Fitzgerald, in which he called his neighbours 'the rudest lot of impertinent, prying people that I ever had the misfortune to live near', was written in September 1889.

6. Rollo Appleyard refers to Heaviside's home as 'a temple of wisdom, the place of the oracle, the court of ultimate appeal' in his book *Pioneers of Electrical Communication*.

7. The statement 'There are no "longitudinal" waves in Maxwell's theory' was the opening line in Heaviside's article 'On Compressional Electric or Magnetic Waves', published in *The Electrician* in November 1897. The article was reprinted as Appendix D to *Electromagnetic Theory*, Volume II.

8. Heaviside asked Planck to clarify his statements on entropy in a letter of 20 February 1903 to *The Electrician*.

9. *The Electrician* printed Planck's reply on 6 March 1903. The fact that the great Planck replied quickly, or indeed at all, shows that Heaviside had established himself as someone to be taken seriously by top scientists, even when venturing outside his own field.

10. Heaviside's letter to Lodge, reporting the loss of his housekeeper, was written in February 1899.

11. The visitor who remembered Oliver using the pianola with 'great vigour' was Mr W.G. Pye, quoted by G.F.C. Searle in *Oliver Heaviside, The Man*.

12. Fleming produced his polemic on the stultifying effect of the Post Office's monopoly in February 1901. I am obliged to Paul Nahin for this quotation. In *Oliver Heaviside: Sage in Solitude* he gives the source as *The Nineteenth Century and After*, Volume 49.

13. Heaviside's suggestion that distortion could be reduced by inserting inductive coils in telephone lines appeared in his article 'Various ways, good and bad, of increasing the Inductance of Circuits', first published in *The Electrician* in November 1893 and reprinted the same year in Chapter 4 of *Electromagnetic Theory*, Volume I.

14. Pupin sang Heaviside's praises in his paper 'Wave Propagation over Non-Uniform Cables and Long Distance Air Lines', published in the *Transactions of the American Institute of Electrical Engineers* in May 1900.

15. The 10th edition of the *Encyclopædia Britannica*, containing Heaviside's article 'Theory of Electric Telegraphy', was published in June 1902. His fee for the article was a little over £15. It was republished in Chapter 10 of *Electromagnetic Theory*, Volume III.

16. Heaviside wrote to Lodge, commiserating with him on Fitzgerald's death, in February 1901.

17. Heaviside's lamentation of the 'brain-wasting perversity' of his countrymen in clinging to outmoded units like inches, feet, ounces, pounds, quarts, gallons and acres was first published in *The Electrician* in 1891 and can be found towards the end of Chapter 2 of *Electromagnetic Theory*, Volume I.

18. The SI scheme took in Heaviside's idea in principle but minimised the disruption to the system of units by defining μ to include the factor 4π (and hence ε to include the factor $1/4\pi$), an idea first put forward in 1901 by the Italian engineer Giovanni Giorgi. See also Note 10 to Chapter 5.

19. P. Hammond refers to Heaviside's removal of the 4πs as a 'cunning device' in his book *Electromagnetism for Engineers*.

20. Heaviside's refusal to provide a synopsis of the prospective Volume III of *Electromagnetic Theory* was expressed in a letter of January 1900 to editor Carter at *The Electrician*. The words are taken from a surviving draft of the letter.

21. The letter to Bethenod in which Heaviside explained why he refused the Hughes Medal was written in early 1921. We know of it from the surviving part of a draft held by the Institution of Engineering and Technology. The draft also reveals that Larmor did his best to help Heaviside. He wrote again after a few years, repeating the offer of the Hughes Medal and enclosing a gift of £100 from the Society. Oliver refused both.

22. On another occasion, when repaying Searle a loan, he wrote:

> Dr S. Te igitur, V.P. Ego te remittare £10 (decem pundit Anglorum in returno lonorum).
> Vos recipe obligato. Vale, Oliverius Heavisidius.

V.P. stood for Vir Preclare (Eminent Man). After Te igitur this was Oliver's favourite Latin expression, taken, as were most of the others, from the citation for his honorary doctorate from Göttingen.

23. In a later notebook entry, Heaviside referred to 'that dreadful malaria I was suffering from at N.A., with internal ruptures'. It is possible that malaria was, indeed, the cause of this and earlier episodes of illness but we cannot be sure.

24. Heaviside told Searle of his move to Torquay in a letter of July 1908.

Chapter 11   A Torquay marriage

1. Heaviside wrote to Searle on 'the great lentil question' in January 1912.
2. The story of a letter addressed to 'Inexh. Cavy. Torquay' arriving safely is reported by Rollo Appleyard in *Pioneers of Electrical Communication*.
3. At this time the Institution of Electrical Engineers had no statue of Faraday, though the Royal Institution had a very fine one. Much later, the IEE had a copy of the RI statue made and installed outside its own headquarters in Savoy Place.
4. Although the electron slotted in neatly in the end, the process of assimilation was far from smooth; there were many false starts and trips up blind alleys. The process started a decade or so *before* J.J. Thomson produced experimental evidence of the electron's existence. Lorentz made the crucial breakthrough in the early 1890s by assuming, in his theoretical calculations, that charged material particles existed in their own right and that their charges were not simply by-products of the field. British Maxwellians were slow to come round to this point of view but eventually Larmor, who had started by modelling electrons as vortices in the aether, found his own theory converging with that of Lorentz. Larmor, an Ulsterman who had settled with ease into the life of a Cambridge academic, was brilliant but sometimes got carried away by grand notions and his own facility for mathematical model-building. Fitzgerald played a key role behind the scenes, encouraging, advising and cajoling Larmor, and he was pleased with the outcome. Reviewing Larmor's book *Aether and Matter* in 1900, he wrote that, 'far from being contrary to the spirit of Maxwell's Treatise', the existence of electrons was 'in reality quite compatible with it, and from some points of view even essential'.

    These were deep waters and most scientific historians would say that Fitzgerald was oversimplifying things. Jed Z. Buchwald gives a scholarly account of the electron's emergence and eventual assimilation in his book *From Maxwell to Microphysics*. Heaviside used the term 'electron' freely in his later papers but what he meant by it was simply a point charge of arbitrary value.

5. Some legends have grown about Heaviside's unpublished Volume IV, stemming principally from the work of H.J. Josephs, a mathematical physicist with the

Post Office. He had been captivated by his first sight of the unpublished papers in 1928 but wasn't able to study them until the late 1940s, by which time they had suffered the effects of wartime storage in a damp cellar. He then attempted to construct some semblance of the fourth volume from odd entries in Oliver's notebooks and hundreds of loose scraps of poor-quality mildewed paper, containing only scribbled mathematics with no accompanying words. In 1949 some more papers were discovered. These, Josephs asserted, held a draft of 'a substantial part of the intended fourth volume' but they, too, were damaged and largely illegible. From such difficult material, Josephs produced a long article in 1950 for *The Heaviside Centenary Volume.* In it, he conjectured that Heaviside had been working on a unified field theory of electricity and gravitation, and gave many examples of Oliver's apparent attempts to bring mathematics to bear in this grand design. Josephs followed up with several more articles but none of his contemporaries found his arguments convincing and it now seems that the idea of a unified field theory was more Josephs' than Heaviside's.

In 1957, more of Heaviside's papers were found under the floorboards of an upstairs room in the Paignton branch of Barclays Bank, formerly Charles Heaviside's music shop. They filled three large sacks. Oliver had lived in the flat over the music shop from 1889 to 1897 and had not thought it worth taking the papers with him when he moved to Newton Abbot. Why did he put them under the floorboards? Presumably to improve the heat insulation in his room.

6. The same note casts a new light on Oliver's brother Charles. He had borrowed £350 from Mary, presumably to help expand his music business, and had paid back £110, but was proposing to keep the remaining £240 to pay for funeral expenses and the cost of winding up her affairs when she died. The immediate effect of the change of ownership from Oliver's viewpoint was that instead of paying Mary £90 a year for rent and board he took on most of the household expenditure directly.

7. We don't know the details but a surviving draft letter shows that he asked an acquaintance called Sir Andrew Noble for a loan of £400 to enable him to complete the purchase of the house, and that brother Arthur, to whom he was already in debt, had agreed to lend him a further £250 to help repay Sir Andrew. The purchase price was probably well over £1000 – Mary's mortgage had been £1500 when Oliver came to Homefield.

8. Heaviside wrote to Searle, complaining, not for the first time, of Mary's unreasonable predilection for cold rooms and draughts, in January 1912.

9. Searle's friend Crowther visited Heaviside in 1914 and recalled the experience in a letter of September 1949 to Searle. The incident bears a remarkable resemblance to a reminiscence of H.M. Butler, who was Master of Trinity College, Cambridge, in the late nineteenth century. He had been a student at the College with Maxwell in the 1850s and remembered an afternoon walk with him along the banks of the Cam. Butler said that he understood hardly a word of his friend's outpouring of his latest thoughts on electricity and magnetism but that he wouldn't have missed it for anything.

10. Alan Heather, a relation of Heaviside, recalls how his mother, Thelma May, used to describe Oliver as 'an awkward old cuss but a brilliant man'. She was the great-granddaughter of Oliver's uncle, George Heaviside.

11. Heaviside had already, in 1908, accepted honorary membership from the British Institution of Electrical Engineers without drama. But the offer from the American Institute of Electrical Engineers was a different matter because, as he saw it, to accept would mean being 'tacked on' to Pupin. The fact that he did accept in the end suggests that, despite his grumbles, he recognised and valued the high regard in which the American engineers held his work.

12. Heaviside's letter to Behrend, saying that he didn't wish to be 'tacked on to Pupin, either fore or aft', was written in February 1918.

13. Heaviside put his message to the censor in a letter of April 1918 to Behrend.

14. Gill also wrote to Oliver Lodge on the matter of getting Heaviside to accept money from the American companies. Lodge took the trouble to draft a carefully worded letter to be sent to Heaviside but it was not used, possibly because Gill thought the best chance was through a personal approach by Searle.

15. Searle reports how he muffed his chance of getting Heaviside to accept financial help from the big American companies in his book *Oliver Heaviside, The Man*.

16. Heaviside's letter to the IEE President, Highfield, inviting him to bring a lady for protection against his 'notorious violence', was written in July 1922.

17. Highfield's account of his visits to Torquay, first to prepare the way and then to present Heaviside with the Institution of Electrical Engineers' Faraday Medal, is quoted by Sir George Lee in his article 'Oliver Heaviside – The Man', published in *The Heaviside Centenary Volume*.

18. Heaviside received, and accepted, the Institution's cheque for £100 in December 1921.

Heaviside's legacy

1. Attenuation was Heaviside's term for the proportion of strength lost by a signal as it travelled along the line. He showed that much of the distortion in early lines happened because the higher-frequency components in the signal suffered greater attenuation than the lower. Rayleigh had used the term earlier but Heaviside adopted it enthusiastically and brought it into general use.

2. Stokes' theorem, sometimes called the theorem of version, is one of the most fundamental in the theory of vectors. It says that the line integral of a vector around any closed curve in space is equal to the surface integral of the curl of the vector over any surface bounded by the curve. Up to Heaviside's time, the standard proof involved heavy mathematics: you wrote down the formula for the components of curl in the x, y and z directions and slogged it out. But Oliver's proof, freely adapted for our purpose, runs on the following lines.

On any surface, draw a closed curve, or circuit. If $\mathbf{E}$ is the electric force vector in that region of space, the total force, or voltage around the circuit in, say, a clockwise direction, will be the sum of the forces in each tiny element of the circuit, in other words the *line integral* of the vector **E** around the circuit.

Draw on the surface a line connecting two points of the circuit. There are now two circuits, both containing the line. The sum of the line integrals of **E** around each of the new circuits, again taken clockwise, will be the same as that around the original circuit because the contributions of the two new circuits along the common line are equal and opposite and cancel one another out.

Even if we draw many inter-crossing lines across the original circuit, making a fine mesh, the sum of the line integrals of **E** around every element of the mesh will still be the same because the contributions from the new lines will all cancel out. This will be true even if we make the mesh elements very small indeed, each enclosing an infinitesimal surface area.

At any point in a vector field, curl is a measure of the vorticity, or swirl, of the field at that point. It is itself a vector, pointing in the direction a right-handed screw would move if it turned with the swirl, and its magnitude is the amount of swirl around an infinitesimal loop surrounding the point divided by the area of the loop, when the loop is positioned in the plane of maximum swirl.

To find the total vorticity over a given surface you take the *surface integral* of the curl vector by adding together, for each of the infinitesimal area elements dS which make up the surface, the component of curl perpendicular to the surface multiplied by dS. In each of our tiny mesh elements the component of curl **E** perpendicular to the surface is, by definition, equal to the line integral of **E** around the element divided by the element's area dS. Rearranging things slightly, the component of curl **E** perpendicular to the surface multiplied by dS is equal to the line integral of **E** around the element.

Adding up the values of dS multiplied by the component of curl **E** perpendicular to the surface for all the tiny surface elements gives the surface integral of curl **E** over the whole surface. But the same total can be formed by summing the line integrals of **E** around all the mesh elements. When this is done the contributions from all the internal parts of the mesh cancel out and the result is simply the line integral of **E** around the original curve, or circuit.

Finally, although we have taken **E** to be electric force, the result is a general one; it will hold for any vector. QED.

Heaviside's proof requires little mathematical expertise beyond an ability to visualise lines and surfaces in space. It deals with the properties of *vectors*, unencumbered by the symbols needed to represent them in a particular coordinate system, and hence gives a direct sense of the physical meaning of the theorem. He included a representation of the result in cartesian (x, y, z) coordinates but this was by way of tidying up once the real work had been done.

Maxwell's proof, by contrast, is couched entirely in cartesian coordinates, employs quite advanced mathematical techniques, and occupies about two and a half pages of his *Treatise on Electricity and Magnetism*.

Stokes set the theorem as a Smith's Prize problem in Maxwell's exam year at Cambridge, but he may not be its originator. When Maxwell later wanted to include it in his *Treatise*, with proper billing, neither Stokes nor anyone else could remember who had thought of it first. Historians subsequently found it in a letter dated 1850 from William Thomson to Stokes, so perhaps it should rightly be called Thomson's, or Kelvin's, theorem.

3.   Heaviside introduced the concept of characteristic impedance, although without giving it a name, in Section 12 of the paper 'On Resistance and Conductance Operators ...' (full title in Note 8 to Chapter 9), published in the *Philosophical Magazine* in December 1887 and reprinted in Article 42 of *Electrical Papers*. His apparently effortless derivation of the impedance of an infinite transmission line appears in Section 11. The impedance turns out to be:

$$Z(p) = \sqrt{[(R + Lp)/(G + Cp)]}$$

where R is the resistance of the line, L the inductance, G the leakance and C the capacitance, all per unit length. If $L/R = C/G$, p disappears from the expression on the right, all frequencies are treated alike and distortion is banished. We then have $Z = \sqrt{(R/G)} = \sqrt{(L/C)}$ and this is the characteristic impedance of the distortionless line – it has the dimensions of resistance and is measured in ohms.

4.   Heaviside published only one paper on electrical machines – 'Magneto-Electric Current Generators' – published in the *Journal of the Society of Telegraph Engineers* in 1881and reprinted as Article 18 of the collected *Electrical Papers*. But his notebook for the same year contains 64 pages of work on electrical machines, covering such topics as multiple coils, self-excitation, dependence of current on the load on the motor, and speed for maximum efficiency. Enough to suggest that he could have made a considerable contribution in this field, had he given more time to it.

5.   One technical point needs a little explanation. When carrying out calculations on AC circuits, engineers generally use not the peak values of the oscillating currents and voltages but their root mean square, or RMS, values. In Britain's supply grid to domestic dwellings the voltage oscillates between plus and minus 325 volts, and the familiar value of 230 is the RMS voltage – the square root of the average value of the square of the voltage over a cycle. Using RMS values is a practical convenience. For example, in resistive circuits like those for heating and filament lighting, the power consumption is proportional to the average value of the square of the current, so if V and I are the RMS values of voltage and current, and R is the resistance, the power consumption is I^2R, or V^2/R or VI, exactly as for a DC circuit.

6.   The symbol i had been used by mathematicians for the square root of -1 since the 1700s. But by the time electrical engineers began to use complex numbers they had already adopted i (or I) as the standard symbol for electric current. To avoid confusion, they decided to use j instead of i for the square root of -1. Curiously, they had originally adopted i (or I) to stand for *intensity* of current – an unhelpful term, now defunct, that Heaviside abhorred. To make his point, he used a different letter, C, to represent current. Everyone now uses i, or I, for current, but the original reason for choosing it has been largely forgotten.

7.   Readers with mathematical training and an interest in the history of operational methods will enjoy the papers of Michael Deakin, of Monash University, who

traces the development of the Laplace transform from its origin in the work of Léonard Euler in 1737. H.S. Carslaw and J.C. Jaeger give a compact account of the transmutation of Heaviside's operational calculus into the modern version of the Laplace transform in the Historical Introduction to their book *Operational Methods in Applied Mathematics*, and the Dutch physicist Balthasar Van der Pol gives an authoritative appraisal of Oliver's methods in the essay 'Heaviside's Operational Calculus' in the *Heaviside Centenary Volume*. Van der Pol points out that Heaviside failed to spot one very useful property of his p functions – the convolution rule. To take an example, the probability distribution of the sum of two independent random variables is given by the so-called convolution of their individual distributions. This is a fairly intricate process but a short-cut method is to take the Laplace transforms of the two distributions, multiply them together and take the inverse transform of the result – a bit like multiplying two long numbers by adding their logarithms and then taking the antilog. Oliver could have sometimes saved himself labour by using his p functions in a similar way.

8. One of the first proper textbooks on Maxwell's Theory, and one of the most influential, was August Föppl's *Einführung in die Maxwellische Theorie*, which came out in 1894. Föppl was Professor of Engineering Mechanics at the Technische Hochschule in Munich. He was a great admirer of Heaviside's work and joined him in believing that a strictly logical 'Euclidean' approach was not the best way to introduce a subject to students.

9. Heaviside's paper 'On Electromagnets, etc.' was published in the *Journal of the Society of Telegraph Engineers* in 1878 and later reprinted as Article 17 of *Electrical Papers*.

10. Heaviside's tribute to Maxwell, already quoted at greater length in Chapter 5, is reported by Rollo Appleyard in *Pioneers of Electrical Communication*.

Bibliography

Appleyard, R., *Pioneers of Electrical Communication*, London, Macmillan (1930).

Arianrhod, R., *Einstein's Heroes: Imagining the World through the Language of Mathematics*, Cambridge, Icon Books (2004).

Baker, E.C., *Sir William Preece, Victorian Engineer Extraordinary*, London, Hutchinson (1976).

Bell, E.T., *Men of Mathematics*, 2 vols., Harmondsworth, Penguin Books (reprinted 1965).

Buchwald, J.Z., *From Maxwell to Microphysics: Aspects of Electromagnetic Theory in the Last Quarter of the Nineteenth Century*, Chicago, University of Chicago Press (1985).

Carslaw, H.S. and Jaeger, J.C., *Operational Methods in Applied Mathematics*, Oxford, Clarendon Press (1941).

Cookson, G., *The Cable: The Wire That Changed The World*, Stroud, Tempus Publishing (2003).

Deakin, M.A.B., 'The Ascendancy of the Laplace Transform and how it Come About', *Archive for History of Exact Sciences*, **44**(3) (September 1992): 265–286.

Fitzgerald, G.F., *The Scientific Writings of the Late George Francis Fitzgerald*, ed. J. Larmor, Dublin, Hodges and Figgis (1902).

Heather, A., *Oliver Heaviside: It's My Genius That Keeps Me Warm*, Paignton, Creative Media (2008).

Heaviside, O., *Electrical Papers*, 2 vols., 2nd edn, New York, Chelsea Publishing Company (1970).

Heaviside, O., *Electromagnetic Theory*, 3 vols., 3rd edn, New York, Chelsea Publishing Company (1971).

Hunt, B.J., *The Maxwellians*, Ithaca, Cornell University Press (1991).

Jackson, W., 'An Appreciation of Heaviside's Contribution to Electromagnetic Theory', essay in *The Heaviside Centenary Volume*, London, Institution of Electrical Engineers (1950).

Jeffreys, J., 'Heaviside's Pure Mathematics', essay in *The Heaviside Centenary Volume*, London, Institution of Electrical Engineers (1950).

Jordan, D.W. 'The Adoption of Self-Induction by Telephony, 1886–1889', *Annals of Science*, **39** (1982): 433–461.

Jordan, D.W., 'D.E. Hughes, Self-Induction and the Skin-Effect', *Centaurus*, **26** (1982): 123–153.

Josephs, H.J. 'Some Unpublished Notes of Oliver Heaviside', essay in *The Heaviside Centenary Volume*, London, Institution of Electrical Engineers (1950).

Knott, C.G., *Life and Scientific Work of Peter Guthrie Tait*, Cambridge, Cambridge University Press (1911).

Lee, G., 'Oliver Heaviside – The Man', essay in *The Heaviside Centenary Volume*, London, Institution of Electrical Engineers (1950).

Lindley, D., *Degrees Kelvin, A Tale of Genius, Invention and Tragedy*, Washington, DC, Joseph Henry Press (2004).

Mahon, B., *The Man Who Changed Everything: The Life of James Clerk Maxwell*, Chichester, Wiley (2003).

Maxwell, J.C., *A Dynamical Theory of the Electromagnetic Field*, ed. T.F. Torrance, Edinburgh, Scottish Academic Press (1982).

Maxwell. J.C., *A Treatise on Electricity and Magnetism*, 2 vols, 3rd edn, Oxford, Clarendon Press (1891; reprinted by Oxford University Press, 1998).

Nahin, P.J., *Oliver Heaviside: Sage in Solitude*, New York, IEEE Press (1988). 2nd edn *Oliver Heaviside: The Life, Work, and Times of an Electrical Genius of the Victorian Age*, Baltimore, The Johns Hopkins University Press (2002).

Pupin, M., *From Immigrant to Inventor*, New York, Charles Scribner's Sons (1924).

Radley, W.G., 'Fifty Years' Development in Telephone and Telegraph Transmission in relation to the work of Oliver Heaviside', essay in *The Heaviside Centenary Volume*, London, Institution of Electrical Engineers (1950).

Russell, W.H., *The Atlantic Telegraph*, London, Day and Son (1866).

Searle, G.F.C., ed. I. Catt, *Oliver Heaviside, The Man*, St Albans, C.A.M. Publishing (1987).

Searle, G.F.C., 'Oliver Heaviside – A Personal Sketch', essay in *The Heaviside Centenary Volume*, London, Institution of Electrical Engineers (1950).

Simpson, T.K., *Maxwell on the Electromagnetic Field*, New Brunswick, Rutgers University Press (1997).

Van der Pol, B., 'Heaviside's Operational Calculus', essay in *The Heaviside Centenary Volume*, London, Institution of Electrical Engineers (1950).

Weightman, G., *Signor Marconi's Magic Box*, London, Harper Collins (1993).

Whittaker, E.T., *A History of the Theories of Aether and Electricity*, London, Thomas Nelson and Sons, 2 vols. (1951, 1953; republished as a single volume, New York, Dover Publications, 1989).

Whittaker, E.T., 'Oliver Heaviside', *The Bulletin of the Calcutta Mathematical Society* (1929), reprinted in Heaviside, O., *Electromagnetic Theory*, Vol. I, New York, Chelsea Publishing Company (1971).

Yavetz, I., *From Obscurity to Enigma: The Work of Oliver Heaviside, 1872–1889*, Basel, Birkhäuser Verlag (1995).

Index

Figure 1   Heaviside's birthplace, 55 King Street, Camden Town. Courtesy of the Institution of Engineering and Technology

Figure 2   The Cart Horse, drawing by Oliver, aged 11. Courtesy of the Institution of Engineering and Technology

Figure 3   *Charles Wheatstone. Courtesy of the Institution of Engineering and Technology*

Figure 4   C.S. Caroline *(the nearer ship) laying the shore end of the 1865 Atlantic cable at Valentia. From 'The Atlantic Telegraph' by W.H. Russell, lithograph by Robert Dudley*

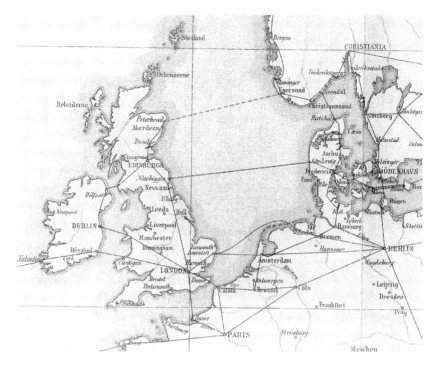

Figure 5   Principal telegraph links in 1869. From a Great Northern Telegraph Company map. Courtesy of GN

Figure 6   The Great Northern Telegraph Company's Newcastle office in the 1870s. The young men on the left are sending and receiving Morse-coded messages using Wheatstone's automatic apparatus, and the boys standing at the table are collecting telegrams for delivery. Courtesy of GN

Figure 7   James Clerk Maxwell. Engraving by G.J. Stodart from a photograph by
Fergus of Greenock. Courtesy of Edinburgh City Libraries

Figure 8   William Thomson (Lord Kelvin). Photograph by Annan of Glasgow.
Courtesy of the University of Glasgow

Figure 9   Oliver Lodge. Courtesy of the Institution of Engineering and Technology

*Figure 10   George Francis Fitzgerald. Courtesy of the Institution of Engineering
and Technology*

Figure 11   *Heinrich Hertz during his military service. Courtesy of the Institution of Engineering and Technology*

Figure 12   *C.H.W. Biggs, Oliver's first editor at* The Electrician. *Courtesy of the Institution of Engineering and Technology*

Figure 13   Josiah Willard Gibbs. Courtesy of Science Photo Library

Figure 14   Sylvanus P. Thompson. Courtesy of the Institution of Engineering and Technology

Figure 15   Oliver Heaviside in his mid-forties. Courtesy of the Institution of Engineering and Technology

*Figure 16   William   Preece.   Courtesy   of   the   Institution   of   Engineering   and
Technology*

Figure 17   Arthur W. Heaviside. Courtesy of the Institution of Engineering and Technology

Figure 18   Oliver, with bicycle. Courtesy of the Institution of Engineering and Technology

Figure 19 *Heaviside family outing at Berry Pomeroy Castle. Oliver, wearing a cap, is standing well behind the group; his head can just be seen beside the pillar between the arches. Others standing, from the left: Lionel (Arthur's son); Sarah (Charles' wife); Isabella (Arthur's wife); Mary Way; Frederick (Charles' son); Rachel; Ethel (Charles' daughter); Thomas; Basil (Arthur's son); Rachel (Charles' daughter); and Arthur. Seated in front are Beatrice (Charles' daughter) and Colin (Arthur's son), with Charles kneeling between them. Courtesy of the Institution of Engineering and Technology*

Figure 20   Michael Pupin. Courtesy of the Science Museum

*Figure 21   Peter Guthrie Tait. Portrait by Sir George Reid. Courtesy of the Royal
Society of Edinburgh*

Figure 22   David Hughes. Courtesy of the Institution of Engineering and Technology

Figure 23   William Burnside. Courtesy of the Royal Society

Figure 24   G.F.C. Searle in his laboratory. Courtesy of the Royal Society

Figure 25   John Perry. Courtesy of the Institution of Engineering and Technology

Figure 26   Homefield, Heaviside's last home. Courtesy of the Institution of Engineering and Technology

Figure 27   John S. Highfield. Courtesy of the Institution of Engineering and Technology

Figure 28   Oliver's grave, before cleaning. Courtesy of Alan Heather

Figure 29   Oliver's grave, after cleaning. Courtesy of Alan Heather

Printed in the USA
CPSIA information can be obtained
at www.ICGtesting.com
JSHW011519221024
72172JS00008B/69

9 780863 419652